STAR HUNTERS

UNITED
STATES

STAR HUNTERS

The Quest to Discover the Secrets of the Universe

DENNIS MAMMANA

Courtesy NASA

A RUNNING PRESS/FRIEDMAN GROUP BOOK

RUNNING PRESS
PHILADELPHIA, PENNSYLVANIA

A RUNNING PRESS/FRIEDMAN GROUP BOOK

Canadian representatives: General Publishing Co., Ltd.,
30 Lesmill Road, Don Mills, Ontario M3B 2T6.

International representatives: Worldwide Media Services, Inc., 115 East Twenty-third Street,
New York, New York 10010.

9 8 7 6 5 4 3 2 1

Digit on the right indicates the number of this printing.

Library of Congress Cataloguing-in-Publication Number 90-52963

STAR HUNTERS: The Quest to Discover the Secrets of the Universe
was prepared and produced by
Michael Friedman Publishing Group, Inc.
15 West Twenty-sixth Street
New York, New York 10010

Editor: Sharon Kalman
Production Editor: Suzanne DeRouen
Art Director: Jeff Batzli
Designer: Kevin Ullrich
Photo Researcher: Gerhard Gruitrooy

Typeset by BPE Graphics
Color separations by United South Sea Graphic Art Co., Ltd.
Printed in Hong Kong by Leefung-Asco Printers Ltd.

This book may be ordered from the Publisher.
Please include $2.50 for postage and handling for each copy.
But try your bookstore first.

Running Press Book Publishers
125 South Twenty-second Street
Philadelphia, Pennsylvania 19103

To The Star Hunters
Of The Past, Present, and Future,
For Their Unending Dedication
To Finding Our Place In The Cosmos.

CONTENTS

The time will come when diligent research over long periods will bring to light things which now lie hidden. A single lifetime, even though entirely devoted to the sky, would not be enough for the investigation of so vast a subject...

"And so this knowledge will be unfolded only through long successive ages. There will come a time when our descendants will be amazed that we did not know things that are so plain to them...

"Many discoveries are reserved for ages still to come, when memory of us will have been effaced. Our universe is a sorry little affair unless it has in it something for every age to investigate...

"Nature does not reveal her mysteries once and for all."

—Seneca, *Natural Questions*
Book 7, first century A.D.

INTRODUCTION

Stand alone on a clear, dark night, far from the blinding lights of a city, and gaze upward toward the starry sky. Who could ever view such a magnificent sight and not brim with awe and wonder?

What are the stars? Who or what is out there? Where did our world come from? And how do we humans fit into the grand machinery we call the cosmos?

We are indeed curious beings. It is our natural curiosity about the unknown—our insatiable desire to know our place in the universe—that links us with our ancestors of ages past, and with our descendents yet unborn. It is, in fact, the single attribute that distinguishes us as humans.

Standing beneath this starry canopy, our earthly concerns and problems dissolve into the darkness. We are no longer a twentieth-century accountant or teacher, plumber or doctor. Instead, we could be a Chaldean shepherd boy tending his

The Omega Nebula in the constellation of Sagittarius is an immense cloud of gas and dust illuminated from within by hot, young stars. Also known as the Horse-shoe Nebula and M17, it is only a tiny portion of a more extensive cloud inside of which new stars are being born.

Courtesy Celestron International

flock, or a brilliant thinker on an evening stroll through the streets of Athens, for the stars we see tonight are the same as those watched by our ancestors ages ago.

In our sky, the hunter Orion still pursues Taurus, the bull. The wandering stars—the planets—continue their intricate dances. And the beautiful, golden face of our moon rhythmically swells and shrinks from night to night, from week to week, just as it has done for countless aeons.

An Age of Discovery

Yet, try as we might, we can never see the sky through the eyes of those who came before us—not because the sky has changed, but because we have. We no longer live in a world of simple shepherds and natural philosophers, a world of celestial heroes and mythological beasts.

Ours is an age of scientific inquiry, an age where fantasy has given way to exploration and fact. And through the ages we have learned that ours is a universe far more complex and beautiful than our ancestors ever could have dreamed.

Today we know our world, our Earth, as a planet: a complex, evolving machine of weather, water, land, and life. We saw it for the first time only three decades ago—a globe of sparkling blue oceans and swirling white clouds, floating peacefully in the vastness of space.

And ninety-three million miles (150 million kilometers) distant lies the brilliant orb from which all terrestrial life draws its breath—the sun. We now know our sun as a hot, glowing sphere of gas, nearly a million miles across. It is a star much like those we see at night. It appears so large and bright only because it lies nearby. Yet, if we could somehow push it trillions of miles away, our sun would shrink and fade until, eventually, it would vanish among the stars of our nighttime sky.

Our Earth journeys around our sun once a year. It is accompanied by eight other planets, dozens of moons, and countless chunks of ice, rock, and metal. Only recently have these distant and mysterious worlds come sharply into focus.

Today we know that the worlds of our planetary family range from heat-seared and cratered Mercury, only thirty-six million miles (fifty-eight million kilometers) from the

In this early sixteenth century fresco by the Italian painter Raphael, God is seen creating the sun and moon. Today, the creation of the planets is believed to have been more scientific in origin.

Left: ***The beautiful ringed world of Saturn was photographed by the*** **Voyager 2** ***spacecraft as it approached the planet on July 12, 1981. By taking photographs through three different-colored filters and combining them into one image, scientists can see details within the planet's cloudy atmosphere.*** **Below:** ***The study of the heavens was considered an art by 1725, when this beautiful celestial sphere was created.***

scorching face of the sun, to frigid and icy Pluto, nearly three billion miles (five billion kilometers) away.

In between lie worlds of remarkable diversity: the infernal heat of cloud-covered Venus; the cold, dry canyons of Mars; the swirling multi-colored cyclones of Jupiter; the glorious rings of Saturn; the strange electro-glow of Uranus; and the magnificent blue clouds of Neptune.

And nearby circle dozens of moons with sights equally astounding: from the exploding volcanoes of Io to the ice sheets of Europa, from the parched craters of Phobos to the nitrogen frosts of Triton.

It is this collection of worlds that we know as our solar system, stretching for billions of miles across space. No matter how hard we try, we can never fully comprehend such distances. The only chance we have is to reduce them to a "human" scale.

Building a Scale Model

Imagine for a moment that we could shrink the sun from its nearly million-mile diameter down to the size of a basketball. The Earth, by

52.
ISTI MIRANT STELLA

Giraudon/Art Resource

This section of the Bayeux tapestry, which measured a total of 230 feet (70 meters) in length, depicts the sighting of a shooting star or some other active nebula. Notice that the characters in the tapestry are pointing towards the sky. At the time this tapestry was woven, the late eleventh century A.D., it was still thought that the Earth was the center of the universe, and all other planets revolved around it.

The moon is our nearest cosmic neighbor, only 240,000 miles (384,000 kilometers) away. Only recently have humans journeyed to the moon and walked upon its dusty soil. We found it to be a silent, lifeless world of craters, mountains, and plains, without even a molecule of air or water to be found.

Courtesy Celestron International

comparison, would be but a pinhead some thirty feet (nine meters) away. The moon would be a grain of sand one inch (three centimeters) from the Earth. Massive Jupiter would be a large marble one city block away, and Neptune a ball-bearing half a mile from the sun.

It doesn't take long to realize how immense our solar system really is. Indeed, the *Voyager 2* spacecraft required twelve years to complete its journey from Earth to Neptune in 1989, and it was traveling at a speed of more than thirteen miles (twenty-one kilometers) per second.

And yet, as immense as it seems, the planetary family in which we live is not all there is, for far beyond lies the realm of the stars.

Across Space and Back Into Time

On a clear, dark night we can see about 2000 stars with the unaided eye alone. With binoculars or a small telescope, millions come into view. Like our sun, stars are hot, blinding spheres of gas. But if our sun were the size of a basketball, some stars would be as large as a house. That they appear so tiny and faint attests to their great distances. Other stars would be the size of a pinpoint by comparison, and are nearly invisible from Earth.

The nearest star beyond our sun is Proxima Centauri, visible to stargazers from Earth's Southern Hemisphere. It lies some twenty-six trillion miles (forty-two trillion kilometers) away (that's twenty-six followed by twelve zeros). If the sun were a basketball in San Diego, the star Proxima Centauri would be a softball—in Maine! And that's the nearest star.

Distances between the stars are so great that measuring them in miles is as ridiculous as measuring distances on Earth in inches. So astronomers measure cosmic distances by the time it takes light to cross them.

Light travels at the incredible speed of 186,000 miles (300,000 kilometers) per second. That's fast enough for a light beam to zip around our Earth 7½ times in just one second. Yet, the distance to Proxima Centauri is so great that light from that star requires 4⅓ years to make the journey to Earth. Since the light that reaches us tonight from Proxima Centauri left that star 4⅓ years ago, we say that it lies 4⅓ "light-years" away. In other words, we see Proxima Centauri not as it is tonight, but as it was 4⅓ years ago.

Even farther are Polaris, the North Star, 8250 light-years away, and Deneb, 1500 light-years distant. And these are still among the nearest of stars. As we peer beyond our cosmic neighborhood, the numbers, sizes, and distances we encounter are even more staggering.

Hundreds of billions of stars, star clusters, and clouds of swirling gas and dust make up our home star system—the Milky Way galaxy. This pinwheel-shaped disk stretches some 600

Courtesy Celestron International

The North American Nebula is named for the continent it resembles. Its gases are lit by hot, young stars that are hidden from view by a lane of dark dust that crosses the nebula. This cloud lies some 2,500 light years from Earth, in the direction of the constellation Cygnus.

million billion (600,000,000,000,000,000) miles from side to side—a beam of light would take a hundred thousand years to cross it.

Ours is not the only such galaxy. Beyond it lie countless others as far as the largest telescopes can see. And the farther we look, the further back into time we see. At the very edge of our visibility, perhaps as far as seventeen billion light-years away, lie the enigmatic quasars. We see these cosmic beacons—perhaps galaxies in the process of formation—as they were when the universe was in its infancy.

New Perspectives

This is the universe we know today—a universe far different than that conceived by the sky watchers of ancient times—far different, even, than that known only a few decades ago. In recent times, we have peered through large telescopes into the depths of space. We have sent robot probes to photograph mysterious worlds beyond our own. And we have taken the first fledgling steps off our home planet to walk the dusty soil of our moon.

Today we view our world, our universe, and our place in it far differently than ever before. Our lives are enriched in ways we may not even imagine—much from the study of astronomy. It has permeated our civilization, and has immeasurably shaped our science, technology, philosophy, religion, music, industry, language, politics, law, art, cinema, and literature. Astronomy gave birth to our calendar, our system of timekeeping, and our ability to navigate the oceans and airways of our world. It has, in fact, made us who we are.

In the modern world, no one can escape its influence. Hardly a week passes that we don't hear of a startling new discovery about the universe—through radio or television, newspapers or magazines. And we consider topics that threaten the cosmic woods: the greenhouse effect, space litter, solar flares.

Indeed, we now view ourselves and our home planet with new and more understanding eyes. Yet, the cosmic perspectives we now share did not come easily. New ideas were often met with skepticism, debate, religious dogma, war, and execution. Sometimes centuries—even millennia—passed before understanding and truth won out, and that only through the undaunted curiosity and determination of countless great scientific minds.

It was Isaac Newton, perhaps the most brilliant scientist who ever lived, who said it best: "If I have seen farther, it is because I have stood on the shoulders of giants." We of the twentieth century, too, stand on the shoulders of giants—star hunters of the past and present—whose perseverance and genius have helped us to understand ourselves as never before, giants who have made the study of our universe a voyage of discovery.

This, then, is the story of these giants. This is the story of the star hunters.

chapter one

THE FIRST STARGAZERS

A gentle wind blows through the brush as darkness falls on the plains of central Africa. The birds have returned to their roosts and are preparing for the night. The animals are quiet now.

The only sound is the soft trickle of a nearby stream. Overhead, the first stars are beginning to appear. And beneath a tree sits a family of apelike creatures, chewing on bones from the day's catch.

The year is 100,000 B.C., and early man is beginning to stir.

People of the twentieth century often think of early man as a grunting, hairy beast with little intelligence, but this is far from the truth. Archaeologists and anthropologists tell us that, since *Homo sapiens* first appeared between 500 and 1000 centuries ago, their brains have changed very little in size and shape. They just used their brains differently than we do today.

Early humans learned that they could perform work more easily with tools made from stone. They could build sheltering structures, hunt for animals, and farm the land. Eventually, they learned to use tools to measure and understand the heavens.

The first humans were hunters, and were more in tune with the cycles of nature than we are today. They watched as the sun rose and set in the sky, and learned that night followed day, and day followed night. They watched as the golden face of the moon waxed and waned month, after month, after month. And through these unending celestial rhythms, they learned to understand the world around them.

They discovered that the appearance of certain stars marked the onset of winter. At these times, the sun appeared low in the daytime sky, and the herds moved on. To escape the cold, early humans were forced to retreat into the safety and warmth of their caves until three full moons passed and new stars appeared at night. Then the sun would rise higher in the sky each day, temperatures would warm, and the cycle of birth, life, and death would begin again.

Day after day, year after year, the cycles became obvious to early hunters, for their very survival depended on them. What also became obvious to them was that the sky seemingly turned daily about a central, flat, and motionless world. That this perfect and untouchable dome of the sky might be the abode of gods who controlled their lives and their world was not long in coming.

Such an idea may have begun one night, while early man was sitting under a tree, resting from the day's hunt. Slowly during the night, the stars disappeared, one by one, until

Above: ***Prehistoric humans spent their time trying to survive. But when darkness fell each evening, they gazed into the sky and watched the glistening stars overhead. And they passed down to their offspring stories about what they saw.***
Right: ***Long before early humans built shelters, they lived in caves and communicated among themselves by carving pictures on the walls. From such carvings, modern scientists have learned that early cave dwellers were exceptional observers of both the land and the sky.***

Giraudon/Art Resource

the sky was overtaken by grayness. Water droplets fell from the sky, and noises rumbled from where the stars once shined.

And then, out of the blackness, a brilliant burst of light reached down and struck the tree under which the hunter was cowering. Rubbing his eyes in disbelief, the hunter might have concluded that whatever, or whoever, was up there was terribly upset with him. This powerful "force" removed the stars from the sky, and its tears were followed by angry bursts of noise and light, which incinerated a tree before his very eyes.

These "sky-gods" now had the hunter's complete attention and respect. And though he didn't understand them, early man vowed to do whatever he could to predict or avoid such angry outbursts in the future.

Who were the first humans to discover the cycles of the heavens and suspect they might foretell events on Earth? Who were the first to gaze at the stars and plan rituals to appease the sky-gods? What fabulous tales did early man tell to explain the mysteries of the sky above? Unfortunately, we will never know, for the answers to these questions are lost in time.

What we do know is this: Our place in the universe had been firmly established. The sky had become the abode of gods and angels, of supernatural powers that controlled events here on our central and all-important Earth. Only by carefully watching the heavens above could early man hope to understand the will of the gods and make efforts to please them.

Humans began their existence as hunters. They quickly learned that finding animals to hunt depended upon seasonal changes, and they learned to predict when these changes would occur by carefully watching the cycles of the celestial bodies.

THE FIRST STAR HUNTERS

It was under the clear, dark skies of the Middle East that the first true star hunters began their work about 6000 years ago.

From towering ziggurats, early Babylonian sky watchers carefully charted the movement of the heavens. Their purpose was not to learn the mechanisms by which the universe worked; they sought, instead, the will and wisdom of the gods.

From the earliest days of Babylonian history, the most powerful and revered of all deities was the sun. It was the provider of

Right: ***Along the banks of the Tigris and Euphrates rivers in what is today known as Iraq, King Hammurabi established the city of Babylon around 1700 B.C. It was here that the first true sky watchers carved their observations of the stars and planets onto clay tablets.*** **Below:** ***Thousands of clay tablets have been uncovered from the region once known as Babylonia. This tablet from about 500 B.C. records the motions of the planet Jupiter. Such motions were important, they believed, because they could help stargazers interpret the will of the gods.***

light and power, the regulator of the seasons, and the giver of life. The sun was accompanied in the sky by a lesser god, the moon, which wandered about from night to night, changing its shape as it moved.

The stars also provided an endless source of inspiration. The ancient sky watchers projected their heroes and demons, and their worldly fears and desires into the stars. And they created the constellations—unchanging pictures from which intricate stories could be woven and passed on from generation to generation.

It was among these timeless figures that the ancients saw other lights moving about in silent, graceful arcs from night to night, from month to month. These were the "wanderers," the planets. Five were visible to the eye: Mercury, Venus, Mars, Jupiter, and Saturn. They seemed to possess special powers, for their bewildering loops might be controlled only by powerful gods.

Stargazers of old discovered that these seven special bodies—the sun, moon, and five planets—did not wander about the sky aimlessly. They seemed, instead, to move only through twelve special constellations, or houses. They called this band of constellations the zodiac, and they believed that the locations of these seven heavenly orbs somehow foretold events here on Earth.

It wasn't long before "astrologer-priests," as they were called, began to interpret the "will" of these celestial gods, and to use it to make predictions of earthly events. Some of the "cosmic connections" they found were real—tides, growing seasons, weather—and reinforced the astrologers' beliefs that the sky-gods controlled the world around them. But other such connections were not real, and when the astrologer-priests tried to use the stars to predict rebellions, wars, sickness, and death, it didn't always work. This, they believed, was simply because they hadn't yet learned to completely interpret the will of the capricious gods.

82
7-14
509
92687

MEANWHILE, IN CHINA...

Around the same time, and halfway around the globe, others were also watching the skies.

Perhaps the greatest observers of the ancient world, the Chinese watched as planets wandered about, new stars flared into view, comets came and went, and eclipses darkened the land. And they carved their meticulous observations into turtle shells and bones for future sky watchers to use.

In their day, the ancient Chinese stargazers identified 300 constellations around the sky. Young stargazers could begin serious work only after they had learned to identify every one of them.

The Chinese knew the celestial cycles well and learned that they could be predicted mathematically. For example, they found that eclipses of the sun and moon repeat every eighteen years.

The rulers of the day were particularly interested in such predictions, for they could use them to gain power over their people. They often hired these stargazers to chart the heavens and advise them on crucial decisions. Court astrologers were powerful men indeed. On their advice often rested the fate of entire empires. Mistakes were not tolerated.

One legend tells of two astrologers named Hi and Ho, who lived during the twenty-second century B.C. They predicted that an eclipse of the sun would be visible from China in the year 2135 B.C.

Hi and Ho were so elated about their prediction that they began to drink rice wine to celebrate. The more they drank, the happier they became, and they forgot to tell the Emperor about the upcoming eclipse.

One day the sky darkened, and terror struck the villages. The emperor was furious that he had missed an opportunity to flaunt his power, and ordered the two drunken astronomers beheaded.

With their lives constantly at stake, astrologers around the ancient world soon learned to hedge their predictions to match any and all circumstances—a practice that continues to this day. A quick check of the horoscopes shows that any of the predictions can safely be made to fit any lifestyle or any personality.

This observatory at Peking, in China, provided ancient Chinese stargazers with the facilities to learn and chart the 300 known constellations.

Snark/Art Resource

North Wind Picture Archive

North Wind Picture Archive

The entire economy of ancient Egypt depended on farming, and thus on the Nile River. The Egyptians watched the sky to develop a system of timekeeping to calculate the length of the year and to learn when to plant and harvest their crops.

THE DAWN OF CIVILIZATION

By about 3000 B.C., the cycles of the heavens were well known, and the first glimmers of civilization were beginning to appear. One of the greatest arose along the banks of the Nile River, in the area known today as Egypt.

The ancient Egyptians were primarily farmers. They knew that each year about the same time the Nile River would flood, signalling that agricultural activity could begin. Since their economy depended upon knowing this cycle with precision, Egyptian stargazers watched and charted the heavens in search of a reliable way of predicting this event.

The sky watchers in the court of King Menes discovered that the Nile would flood each year as the brilliant star Sirius appeared on the eastern horizon at dawn. They began to watch every year for the early-morning rising of Sirius, and learned to predict with remarkable accuracy when flooding would occur and farming could begin.

Being tremendously practical people, the Egyptians used this celestial cycle to develop a calendar with 365¼ days. They divided each day into twenty-four hours, and accurately measured time by sundials and water clocks. And if that wasn't enough, they used the stars to plan and build temples and pyramids, to develop mathematics that could perform intri-

Right: ***The Egyptians depended upon the regular flooding of the Nile to irrigate their crops. By studying the stars, they could better predict this yearly occurrence.*** **Below:** ***The Egyptians also believed that the sun embarked on a daily journey across the land, thus providing an explanation of why it arose each morning and set each evening.***

North Wind Picture Archive

North Wind Picture Archive

Giraudon/Art Resource

This Ptolemaic zodiac, now in the Louvre Museum in Paris, France, maps the constellations Ptolemy built upon Aristotle's system.

cate calculations, and to build dams and ditches to irrigate their crops.

Yet, when it came to understanding the universe and their place in it, the Egyptians relied more on traditional religious practices than on observations and calculations. They developed elaborate schemes to explain the cosmos and rejected all ideas that didn't agree with holy doctrines.

The ancient Egyptians believed that the universe was shaped like a rectangular box, with its longer sides running north and south and the flat earth lying on its floor. The heavens were formed by the body of a goddess named "Nut," who stood over them supported only by her knees and elbows. Around their world flowed the great celestial river Ur-nes, on which sailed boats that carried the sun, the moon, and the planets on their daily jaunts across the sky. And they believed the stars to be lamps carried in the hands of deities.

Before the fifteenth century B.C., this strange combination of careful, accurate observation and unscientific interpretation was the rule rather than the exception. This, however, would soon begin to change. Only a few hundred miles northwest of the Nile River Valley a new and powerful civilization was taking hold, one that would change forever our perceptions of the cosmos.

© Gian Berto Vanni/Art Resource

chapter two

EARLY THINKERS

The Parthenon is perhaps the greatest symbol of ancient Greek civilization; its geometric order echoed the Greek desire to find a rational explanation behind natural occurrences.

It was along a jagged, mountainous peninsula in the Mediterranean that another culture sprang up—one which gave us poetry, literature, philosophy, and mythology. It was a culture that provided a view of the universe which thrived for more than 1500 years. It was one of the greatest civilizations of all time—ancient Greece.

In many ways the early Greeks were much the same as those who came before. They, too, watched the heavens and charted the movements of the sun, moon, and planets, and used these discoveries to regulate their daily activities.

But the Greeks were also very different. Their philosophers were not tied to a religion that limited their intellectual inquiry. Unlike the Babylonians, who attributed motions of planets to capricious gods, the ancient Greeks were interested in finding natural causes. And rather than try to decipher the will of the gods and answer questions for

North Wind Picture Archive

The civilization of ancient Greece flourished on a mountainous peninsula crossed by narrow valleys and long gulfs. Its potential for farming drew Hellenic tribes from the north who established colonies on the Aegean Islands, the Ionian coast of Asia Minor, southern Italy, and Sicily.

anxious kings, they worked to increase human understanding of the world and universe through observation.

New Directions

It was in the sixth century B.C. that things started to change in the ancient world. In the East, philosophers such as Confucius, Buddha, and Lao-Tsu were founding schools of thought that would forever change the course of human philosophy. And in Greece, a "fever" of learning took over—a great desire to find rational explanations behind natural occurrences.

The Greeks had come to believe that the universe was understandable because it had internal order—rules that it must obey. Many thinkers devised ways to explain the celestial sights, while others were interested in learning the physical and chemical makeup of the universe.

Perhaps the first of all true astronomers was Thales. In his study of the world and the universe around him, Thales developed a method of building theoretical models to explain the structure and behavior of nature, much as we do today. And he constantly sought confirmation of his ideas through analogies with more familiar events.

Thales believed that the Earth on which he lived was a circular disk afloat on a great sea of water. But, unlike those who came before him, Thales reasoned that the land was born from the oceans by material forces, not created by the hands of powerful gods.

Today we know that his descriptions of the world are incorrect, but Thales took a major first step toward true scientific inquiry when he proposed a natural, rather than a divine, origin of the world.

This was a time of great thought, and Thales was not alone in trying to solve the mysteries of the universe. With little to work with besides their eyes and their minds, the Greek philosophers came remarkably close to the ideas we have today.

One such philosopher was a young contemporary of Thales named Anaximander of Miletus. Anaximander was bold enough to propose that the Earth was suspended freely in space. He suggested that it was surrounded by turning domes, each with holes that allowed light from heavenly fire to shine through and create the stars, the planets, the sun, and the moon. He was the first to estimate the sizes of the sun and moon and was

Historical Picture Service/FPG International

also the first person in Greece to make and use a sundial.

A slightly different approach was taken by another philosopher of the day, named Anaximenes. Anaximenes considered the flat Earth to be floating on air. The light and heat of the sun, he believed, were created by friction as the sun moved through the air. The stars and planets, on the other hand, were created by moisture rising from the ground and catching fire once in the sky. No one felt their heat because they were much too far away, farther even than the sun or the moon.

And then there was Empedocles who devised four elements—earth, water, air, and fire—which he believed made all things. He

Left: ***While the Greek civilization rose in Europe, great thinkers such as Buddha were founding schools of rational thought in Asia.*** **Inset:** ***This vase painting from ancient Greece depicts a typical school day. Students listened to great thinkers lecture about such diverse fields as poetry, mathematics, music, and natural philosophy.*** **Below:** ***This drawing of Raphael's*** **School of Athens** ***portrays the great thinkers of the day gathered around Aristotle and Plato.*** **Overleaf:** ***The Acropolis was built at the height of Athenian power. Like its art and architecture, Greek thought has survived dark periods when it was lost to us.***

North Wind Picture Archive

TIMEO
ETIC

viewed the universe as a crystalline sphere, and held that the atmosphere was a material substance, not a void. Many of these thinkers worried about the material make-up of the universe, but others were busy exploring the importance of numbers to deduce order around them.

It was Pythagoras, the great geometrician, who first stated that the Earth was a sphere, and that the planets moved independently through the sky. He considered the cosmos a finely tuned musical instrument, and believed that the planets, sun, and moon emitted tones as they moved about their orbits. He also believed that the distances of the planets from the centers of their orbits should fit precise mathematical (and musical) laws.

While Pythagoras made history's first effort to unify concepts of the universe—to show how all its components interacted to create the universe people observed—things soon became extremely complicated and confusing. For example, Eudoxus of Cnidus calculated a system of twenty-seven concentric spheres, which he believed were necessary to carry the planets around the Earth and account for their mysterious looping movements in the sky. And the Pythagorean thinkers soon applied their skills to other fields of study, such as politics, art, and philosophy—realms that even today cannot be understood by so impersonal a language as mathematics.

THE GREAT SYNTHESIZER

Toward the middle of the third century B.C., a remarkable man came onto the scene. His name was Aristotle. Aristotle was a Greek thinker in the truest sense of the word. He became known throughout the land as an expert on philosophy, history, politics, ethics, poetry, drama, and biology.

Aristotle did what no other philosopher had been able to do. He melded the ideas of his predecessors—the four elements of Empedocles, the mathematics of Pythagoras, the concentric spheres of Eudoxus—into one great conceptual model of nature. It was, indeed, a grand unified theory that explained the workings of the entire known universe as never before.

The Universe of Aristotle

Aristotle's view of nature was based primarily on the inherent tendencies of matter. He believed that the universe was composed of four elements—earth, water, air, and fire—and that each had a natural resting place. Earth and water were heaviest and naturally moved downward, while air and fire were light and so always moved upward.

Aristotle reasoned that the Earth must be a sphere, for if all material fell downward, the only shape it could produce was a sphere. As proof, he showed that the sky doesn't appear the same from different parts of the world as it would if the Earth were flat. And, he explained that the only shape that could always produce a circular shadow on the moon during a lunar eclipse is a sphere.

Opposite page: ***This detail of Raphael's*** **School of Athens** ***shows Plato*** **(right)** ***and Aristotle*** **(left)*****, two of the most brilliant minds, ever.***
Left: ***Aristotle's ability to synthesize others' ideas led to the view that the earth was a solid, immovable body around which the sun, moon, planets, and stars all turned on immense crystalline spheres.***
Below: ***Pythagoras believed that everything in the universe was governed by musical (and mathematical) laws, and spent his life trying to discover them. He was perhaps the first thinker to develop a unified, mathematical vision of the heavens.***

Above: ***Residents of the ancient city of Athens passed by the Tower of the Winds daily. Carved around the top of the tower were a series of figures that represented the prevailing winds. The Triton on top was a bronze weather vane that pointed its wand toward one of the carvings.*** **Below:** ***The star cluster known as the Pleiades was named by the ancient Greeks for Atlas, the titan who held up the heavens, and his daughters. The cluster can be found in the constellation of Taurus, the bull, and is sometimes called the Seven Sisters.***

Because of the natural tendency of heavy elements to fall downward, Aristotle reasoned that the Earth must be at the center of the universe, surrounded by an ocean of water, a sphere of air, and a sphere of fire. The Earth, he believed, was stationary, for no object this heavy could possibly move.

Revolving about this immovable Earth each day were seven crystalline spheres that carried with them the sun, moon, and five planets. Surrounding them was the sphere carrying the fixed stars.

Aristotle believed that the Earth was corruptible and changeable, as anyone could plainly see from day to day, but that the farther away the celestial spheres were from us, the more perfect they were. To prove this idea, he pointed to the moon. He explained that its shadings and imperfections proved it must be nearest to us, while the sun, planets, and stars—none of which showed imperfections—must be farther away.

Beyond the spheres of the planets, sun, moon, and stars, Aristotle proposed the most perfect of all spheres—the Primum Mobile, the First Mover. It carried no bodies, but was instead the master gear that set in motion all other spheres.

At the Same Time . . .

Another philosopher busy trying to explain the cosmos was Aristarchus of Samos. What made Aristarchus different from Aristotle was that he combined the reason of Aristotle with meticulous observation and mathematics.

Aristarchus was the first to propose that it was the sun, and not the Earth, that dominated the center of the universe, and that our world circled it once a year. He explained that the daily rising and setting of the heavenly bodies was caused not by the motion of Aristotle's celestial spheres, but by the twenty-four-hour rotation of the Earth itself.

By methods of triangulation Aristarchus calculated that the sun was much farther from the Earth than the moon, and was also much larger. He also showed that the fixed stars don't appear to change as the Earth moves

North Wind Picture Archive

Perhaps the greatest observer of antiquity was Hipparchus of Nicaea. He calculated sizes and distances for the sun and moon, and measured the length of the year to within 6.5 minutes. Unfortunately, instruments such as those depicted here didn't exist during the time of Hipparchus.

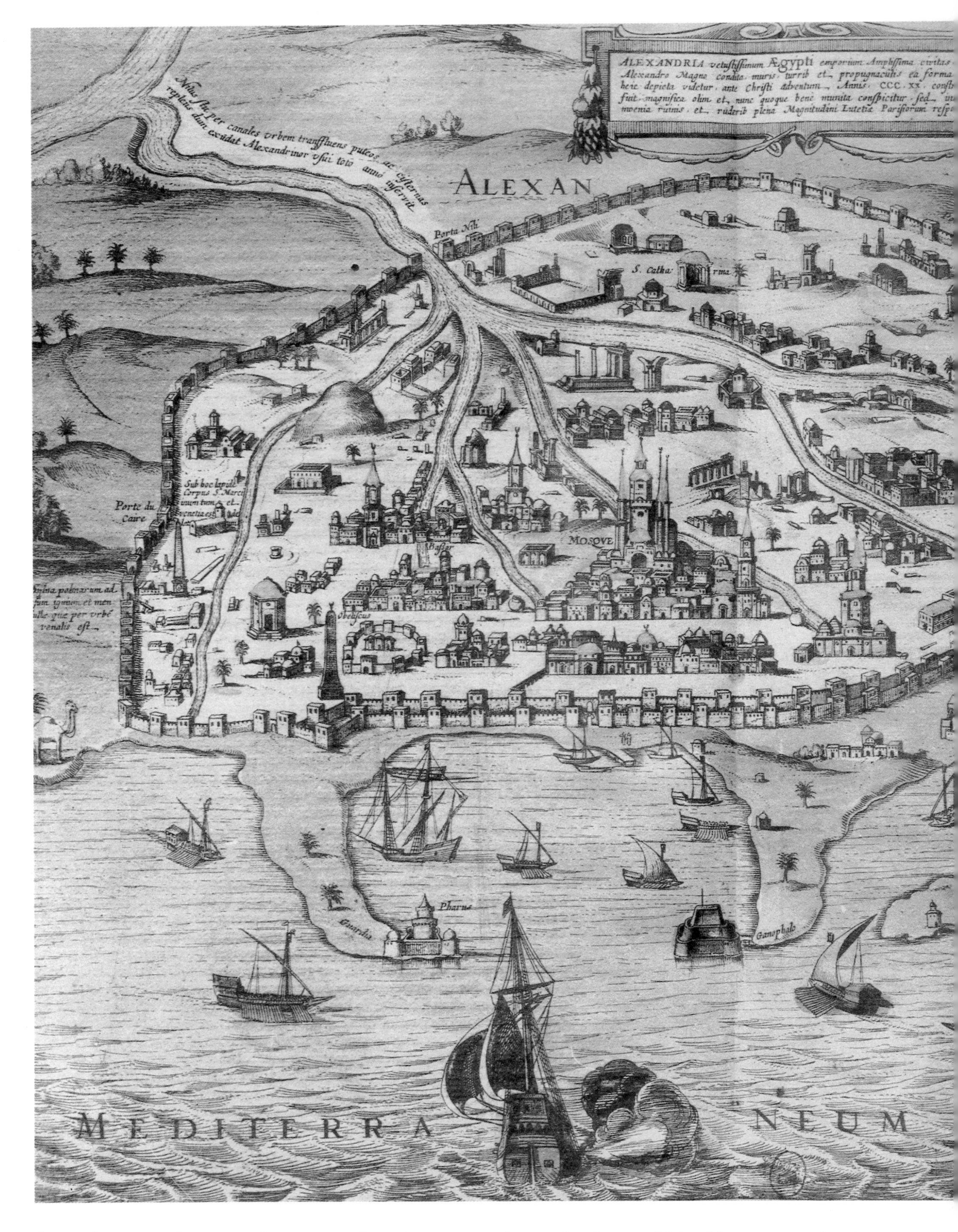

Right: ***The city of Alexandria was a marvelous center of knowledge, commerce, and culture. Scholars and merchants from around the ancient world came to visit its zoos, gardens, and magnificent library. It was here that some of the most remarkable theories of the universe were born.*** **Opposite page:** ***Claudius Ptolemy is often called the greatest astronomer of antiquity. Using his own astronomical observations, as well as those of Hipparchus and the Babylonians, Ptolemy tackled major problems facing thinkers of the day, including the motions of the sun, moon, and planets.***

through space, because they are unimaginably far away.

The systems of both men had serious flaws, and philosophers of the day were understandably concerned. Aristarchus asserted that the concentric spheres of Aristotle didn't explain why the planets moved as they did through the sky, or why some appeared to change their brightnesses over time. Aristotle argued that if the Earth moved at the tremendous speed proposed by Aristarchus, everything and everybody would be blown away by tremendous winds.

Which view of the cosmos was correct, no one knew. Yet, as the Greek thinkers argued amongst themselves, a new and diverse culture was beginning to thrive back in Egypt—one in which scholars from around the world would gather to ponder the mysteries of the universe.

Its name was Alexandria, a spectacular city of marble buildings, wide boulevards, and

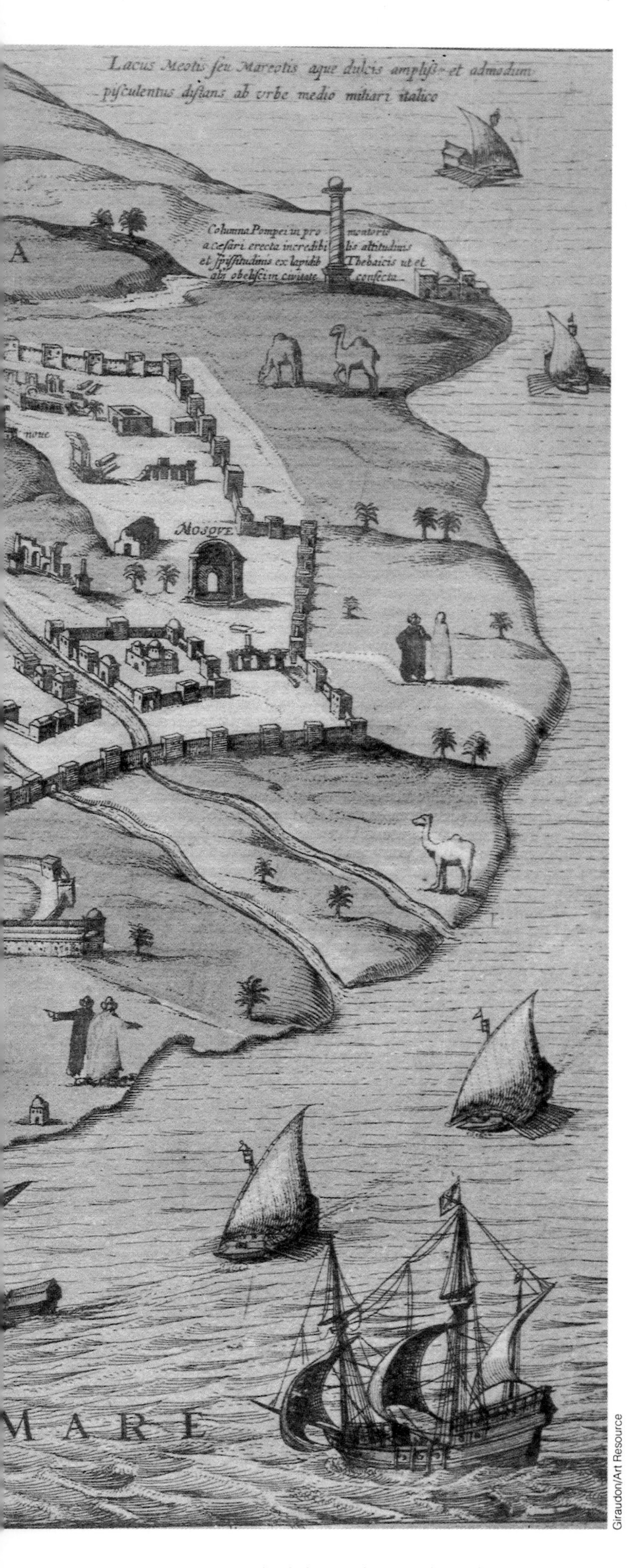

Giraudon/Art Resource

statues. Founded by Alexander the Great to encourage the pursuit of knowledge, Alexandria quickly became a center of commerce, culture, and learning. Here, men and women from all over came to exchange merchandise and ideas: Greeks, Egyptians, Arabs, Syrians, Hebrews, Persians, Nubians, Phoenicians, Italians, Gauls, and Iberians.

At the heart of the city stood a magnificent library. It contained ten large research halls, spectacular fountains and colonnades, botani-

North Wind Picture Archive

cal gardens, a large zoo, and an observatory. On its shelves were handwritten papyrus scrolls, half a million volumes by some of the greatest thinkers who ever lived. It was here that scholars from around the world came to explore such diverse fields as physics, literature, medicine, astronomy, and mathematics.

Aristotle's Defender

One of those was a man named Claudius Ptolemaeus, widely known as Ptolemy. We know very little about Ptolemy's life before he arrived at Alexandria. Even his nationality is uncertain, but his name suggests that he might have had Greek and Roman ancestors.

Ptolemy was a brilliant and prolific author, having written books on music, mechanics, geometry, optics, and geography. He made the first truly scientific maps of the Earth and the heavens, named a number of stars and listed their brightnesses, and compiled some forty-eight constellations that had been handed down from the Egyptians, Babylonians, Greeks, and Romans.

But his greatest work—perhaps the greatest astronomical work of antiquity—is known as *Almagest*. In its thirteen sections, or "books," Ptolemy outlined the state of the astronomical knowledge of the day, and took on the challenge of explaining the irregular planetary motions and brightnesses that had confounded sky watchers for ages.

Overleaf: ***In addition to his work in astronomy and mathematics, Ptolemy was also a great geographer and wrote an excellent book entitled*** **Geography.** ***In it he listed the latitudes and longitudes of many places, described a number of land areas, and described how to construct maps.***

CAVRVS·CHORVS·VEL·IAPIX·SIVE·ARGESTES
CIRCTVS VEL·TRESIIAS
SEPTEN
FAVONIVS·ZEPHIRVS
mare glaciale
EVROPA
DATIA
Pontus euxinus
Sinus hesperus
LIBIA·INTERIOR
AFFRICA
Circulus equinoctialis
Ethiopia sub egipto
ARABIA·FELIX
Barditi montes
ETHIOPIA·INTERIOR
Terra incognita secundum ptholomeum
9 10 14 20 24 30 34 40 44 50 54 60 64 70 74 80
AFRICVS·VEL·LIBS
LIBONOTVS·EVROAV·STER

AQVILO·VEL·BOREAS
CECIAS·APELIOTES
SVBSOLANVS
ASIA
Sacharum regio
Scithia extra imaum montem
Seica regio
Emodi mons
Tropicus cancri
India ultra gangem fluuium
India intra gangem fluuium
Sinus gangeticus
Sinus magnus
Sinarum regio
MARE·INDICVM
PRASODVM
MARE
Terra incognita secundum ptholomeum
Gradus longitudinis ab occidente in orientem
EVRNOTVS
VLTVRNVS·EVRVS

To do so, Ptolemy began with the facts as he saw them. The Earth was round and was firmly situated at the center of the universe. Around it, the heavens turned in perfect circles—first the moon, then Mercury, Venus, the Sun, Mars, Jupiter, and Saturn.

From this knowledge, and a keen desire to save the appearance of Aristotle's system, Ptolemy combined circles upon circles until he devised a method that accomplished his goals.

What he came up with was a complicated series of "epicycles" and "deferents." In this system, planets turned on little wheels as they orbited the Earth on larger ones, much like a Ferris wheel carriage moves in small loops around a bigger wheel. Ptolemy reasoned that, if the universe behaved like this, then an observer on the ground would indeed see planets make looping maneuvers in the sky, and would detect brightness changes since the planets alternately get closer and farther from the Earth.

When it was perfected, Ptolemy's system was horrendously complicated, yet it accomplished his goals. Not only could it explain the irregular motions of the planets and enable sky watchers to accurately predict their positions, but it saved the ideas of the great Aristotle as well.

Why Ptolemy chose to save these ideas is unclear. Perhaps it was Aristotle's antithesis between the corruptible Earth and the perfection of the heavens that reassured a frightened world of the order and permanence in nature. Or, perhaps, it was that Aristotle's system was simply more "obvious" to the average person.

Whatever the case, the systems of Aristotle and Ptolemy flourished, and the ideas of Aristarchus, now known to have been correct, faded silently away.

The oldest known astronomical device is an armillary sphere. This one, based on the Earth-centered model of Ptolemy, shows circles representing the paths of various celestial objects. Armillary spheres were used for observations by such notables as Eratosthenes, Hipparchus, and Ptolemy.

Times Were Changing

Ominous winds of change were beginning to blow. The Catholic Church, which became the most powerful entity of the Middle Ages, approved of Aristotle's and Ptolemy's work and declared that any other belief was an unforgivable and punishable sin.

At the same time, something was going dreadfully wrong in Alexandria. The knowledge that had once been cherished and prized was now being used for the development of weapons, for superstition, and for the amusement of kings.

It was in a moment of great fear and terror that mobs of angry citizens destroyed the great library and all that it symbolized. When they arrived, the last great physicist, mathematician, and astronomer of the age was working inside. Her name was Hypatia, a beautiful woman who had accomplished much at a time when women were considered mere property.

Despite the ensuing attack, Hypatia steadfastly continued to work—until the hoards burst in and dragged her off to be skinned and burned alive.

In one final burst of madness, the crowd returned to burn to the ground the last remaining pieces of the library.

Left: ***Medieval thinkers believed that the positions of the planets and stars could predict events on Earth, and that metals like copper could be changed into gold. The fields of astrology and alchemy were considered important enough to teach in the schools of the day.*** **Below:** ***Around the second century A.D., the Catholic Church became a formidable force throughout Western Europe. Independent scientific inquiry was discouraged, and intellectual curiosity went into hibernation. The Dark Ages had begun.***

An Age of Darkness

In just 800 years, the first real scientific study of the universe had begun, reached its pinnacle, and ended. Novel ways of looking at nature had been developed, and scientific principles to explain it had freely flowed. It was truly a golden age of astronomy, an age like none the world had ever seen.

But now it was over. The Church had gained total power and began dictating its doctrine to the masses. Savagery and violence overtook Western Europe during the fifth century and continued for a dozen generations—a time now known as the Dark Ages. Learning and discovery, once cherished and prized, grinded to a halt. Now, all one could do was gaze toward the sky and wonder—sometimes in admiration, often in fear. This first marvelous age of learning had passed away.

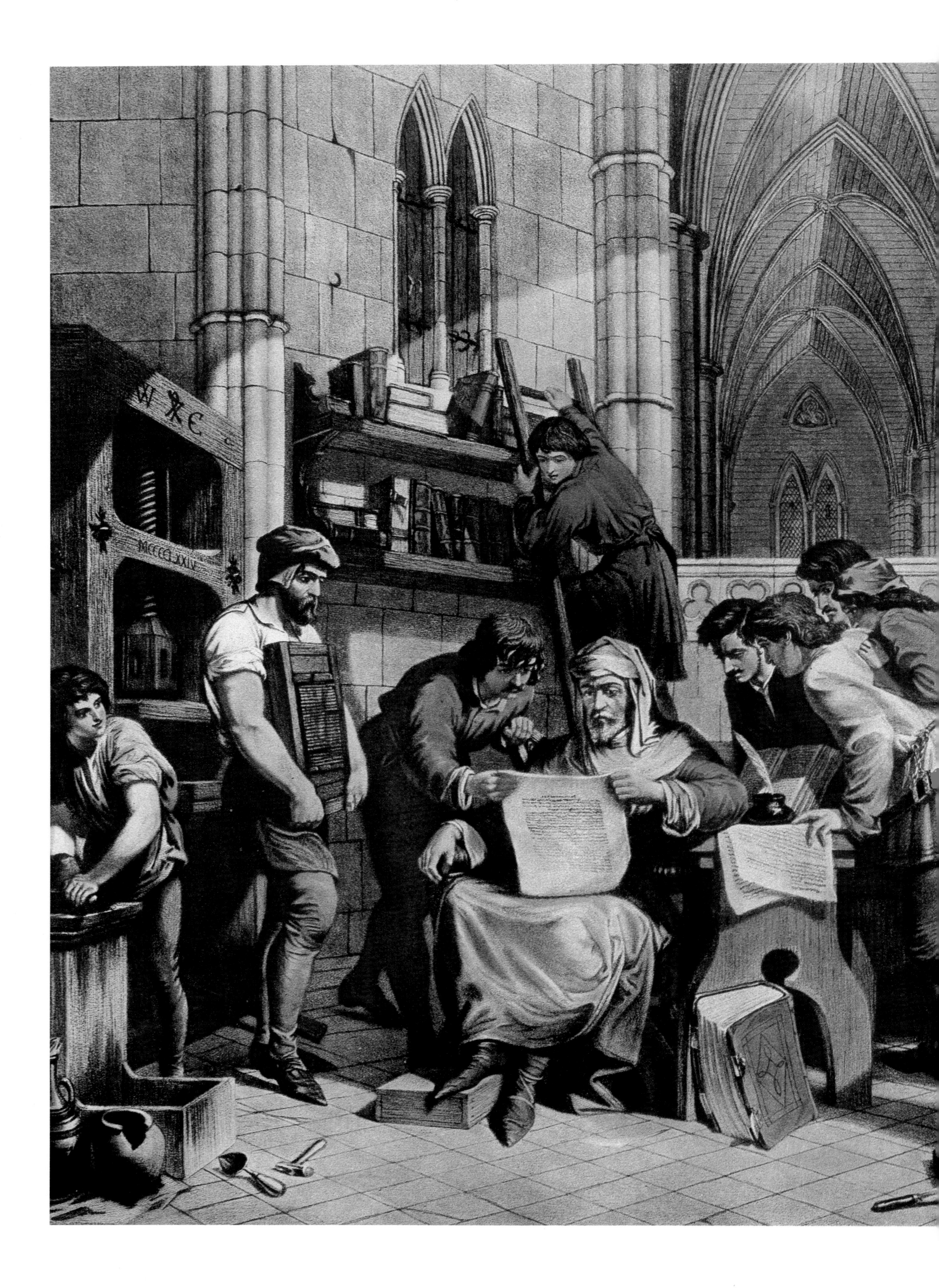
W C
MCCCCLXXIV

North Wind Picture Archive

William Caxton examining the first proof sheet from his printing press in Westminster Abbey in 1475. By the end of the fifteenth century, more than one thousand such presses existed around Europe, and twelve million books had been published.

chapter three

A REAWAKENING

Fifteen hundred years had come and gone. The Earth-centered universe of Aristotle and Ptolemy had become so deeply embedded into human consciousness that to even suggest it was wrong was unthinkable.

But times were changing. From the intellectual darkness of the Middle Ages, a spark of curiosity was beginning to kindle. What caused it is unknown. Maybe it was the invention of the printing press that exposed the public to ideas and questions that contradicted traditional religious dogma. Or, perhaps it was the magnetic compass and Jacob's staff that enabled explorers to venture far from coastlines in search of adventure, fortune, and knowledge.

Whatever it was, it was real. People were once again asking questions, and scientists were beginning to wonder why they should abandon all reason and accept a central and motionless Earth on faith alone.

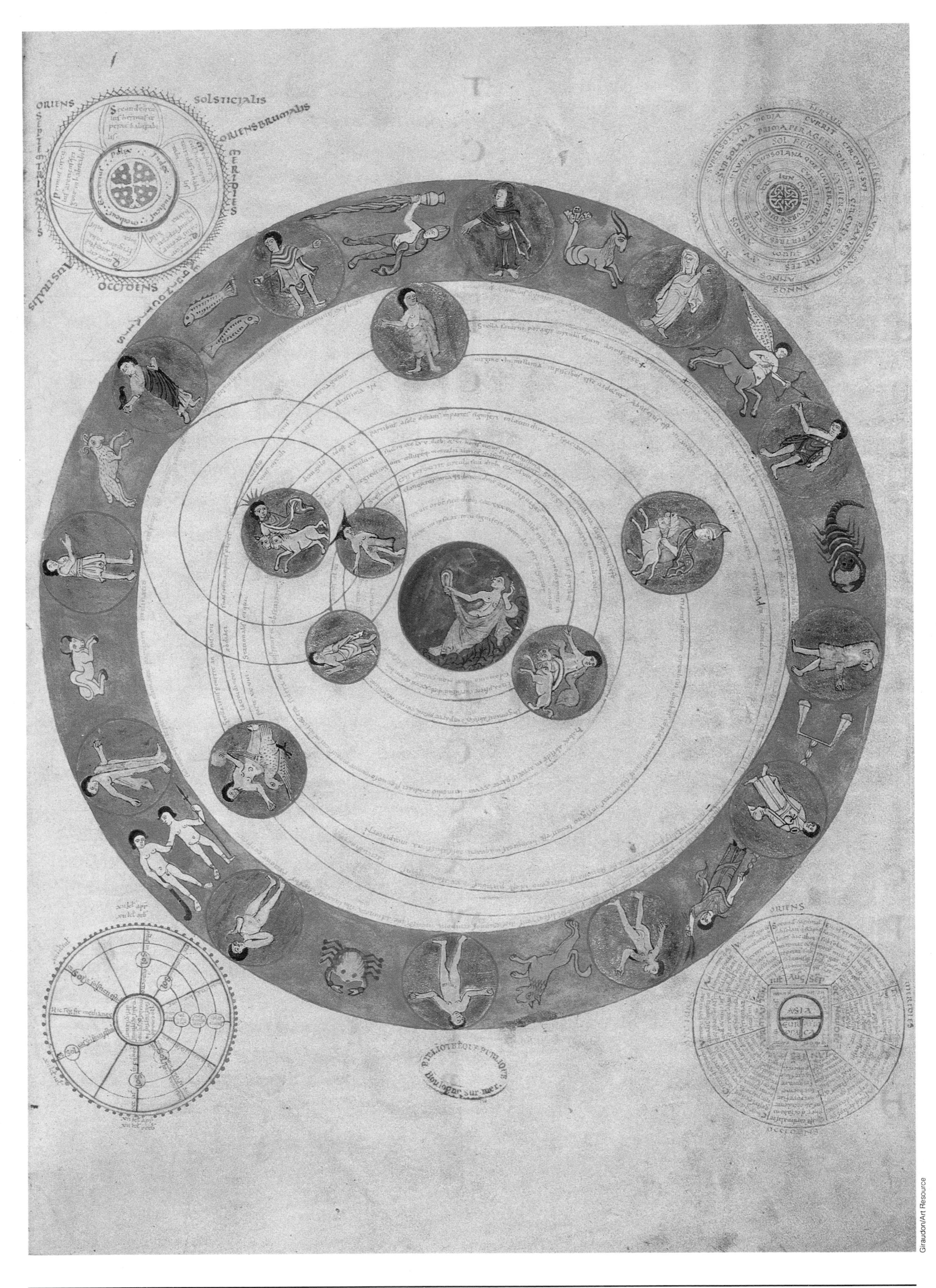
ORIENS
SOLSTICIALIS
ORIENS BRUMALIS
MERIDIES
OCCIDENS
AUSTRALIS
ORIENS
ASIA
BIBLIOTHEQUE PUBLIQUE
Boulogne sur mer

Scala/Art Resource

The Bettmann Archive

Pico della Mirandola* (far left), *the famous Renaissance humanist, helped to spark a major revival of the ideas of Plato and Pythagoras. At the Platonic Academy in Florence, he taught that the Earth-centered universe works as a finely tuned musical instrument.* Left: *It was the Polish astronomer Nicolaus Copernicus who bucked astronomical traditions by proposing that it was the sun and not the Earth which occupied the center of the universe, and that the five visible planets orbit around it.* Opposite page: *This fresco depicts the Medieval view of the zodiac. During this time, also known as the Dark Ages, learning and discovery came to a screeching halt under the auspices of the Catholic Church.

The Italian humanist Pico della Mirandola felt the answer was clear. In 1489, he wrote that the Creator told the first man on Earth: "I have placed thee in the center of the world that thou mightest the easier see that which is about thee . . . " But little did he know as he penned those words, that in only a few decades, a young canon of the Church would propose a theory that would overthrow the wisdom of the ages, and forge a scientific revolution the likes of which the world could never imagine. The young canon's name was Nicolaus Copernicus.

THE REVOLUTIONARY

Copernicus was born on February 19, 1473 in the Baltic port-town of Torun, Poland. After his father died in 1483 Nicolaus was raised and educated by his Uncle Lukasz, an important bishop in Poland.

After studying church law and medicine at major universities throughout Europe, Copernicus became a canon of the Church at age twenty-four. It was there that he was exposed to the new ideas of the Renaissance and developed an intense interest in what he called "the most beautiful of sciences"—astronomy.

During his time in school, young Nicolaus read the works of the ancient Greeks, and found Ptolemy's theory much too complicated to explain the motions of just seven bodies. No, he liked instead the simple elegance of Aristarchus who proposed that it was the Earth, and not the sun, that moved. Someone had to be right, but who?

In 1506, he returned home to Lidzbark Castle to become the private physician of his uncle and, in his spare time, began his struggle to find the right answer. For six years he read and worked. The more he did, the more convinced he became that the motions of the heavens could be explained more easily if it were the sun, and not the Earth, that was motionless.

The Theory

Copernicus looked at it this way. The sun lay in the center of the universe, and orbiting it in giant circles were the planets. The farther a planet was from the sun, the longer it took to go completely around it. Mercury was the closest planet to the sun and took only three months to go around once, while the farthest known planet, Saturn, required thirty years to make its journey.

To Copernicus, the Earth was simply one of the planets. Not only did it rotate once every twenty-four hours, giving rise to the familiar day/night cycle, but it made a complete circle about the sun in just one year. The strange looping motions that some of the planets showed in the sky, he argued, arose simply because the world periodically rushed by certain planets and made them appear to stop

their forward motion and go backwards for a time.

And the stars, said Copernicus, must be globes of fire tremendously far away, for if they weren't, they, too, would seem to wobble during the Earth's multi-million-mile journey around the sun.

It was so simple, he thought. How such an idea could go unnoticed for so long was troubling to him. He knew that to contradict such great philosophers as Aristotle and Ptolemy, not to mention the teachings of the Church, was an unforgivable sin. Still, he concluded, his system was more aesthetically pleasing than those of Ptolemy and Aristotle.

In 1512, while devising a more mathematical treatment of his ideas, Copernicus wrote a small handwritten booklet entitled *Commentariolus* and distributed it among his scientific friends and scholars throughout Europe.

That same year, Uncle Lukasz died, and Nicolaus moved from Lidzbark Castle to the northwest tower of the ancient cathedral at Frombork. He described his new home as the "remotest corner of the Earth," yet it was there that he completed his greatest work.

By 1530, he had finished his book entitled *De Revolutionibus Orbium Coelestium (On the Revolutions of the Celestial Orbs).* Word of his radical theories had spread throughout Europe, and people wanted to know more.

But Copernicus still had some doubts. His work hadn't even been published yet and already he had been criticized for it. One of the earliest attacks came from Martin Luther, who wrote: "Mention has been made of some new astrologer, who wants to teach that the Earth moves around, not the firmament or heavens, the sun and moon.... This fool seeks to overturn the whole art of astronomy. But as the Holy Scriptures show, Jehovah ordered the sun, not the Earth, to stand still."

And as far as the Catholic Church was concerned, there could be no argument. So, to make absolutely certain his theories agreed with observations before he again contradicted the Church, he checked and rechecked his numbers.

North Wind Picture Archive

A very religious man, Copernicus **(left)** ***performed some of his work in this room at the Cathedral at Frombork*** **(opposite page).**

***Martin Luther, a rebel himself, was nonetheless quick to criticize the radical theories Copernicus set forth in* On the Revolutions of the Celestial Orbs.**

The Book

Then, in 1539, a young professor named Joachim Rheticus came to see the aging astronomer to learn of his new theories and to encourage him to publish his ideas for the benefit of all mankind. Still, Copernicus wasn't sure he was ready.

Finally he relented, and in 1543 the book was published. But Copernicus never had a chance to read his own book, for only months before he had suffered a stroke, which left him weak and bedridden. Legend has it that when the first copy rolled off the press on May 24 and was rushed to his bedside, Copernicus looked up and weakly touched the book. A few hours later, he died.

Perhaps it's just as well he never opened the book, for if he had, he would have been furious. The publisher, in an effort to soften the revolutionary impact of the book, had written an anonymous preface describing the work as merely a means of performing more accurate astronomical calculations and not a true expression of reality.

Yet, even with this disclaimer, the book caused quite an uproar. In it Copernicus undermined the authority of the Holy Bible, removed the Earth from its time-honored position at the center of the universe, and single-handedly set the world in motion.

In 1616, the Catholic Church placed the work on its list of forbidden books, a dubious honor that wasn't lifted for nearly two centuries. But before Copernicus' book was banned, another book was published—this by a young, unknown, German math teacher. He wrote that Copernicus was wrong, not because of his theory, but because he used the wrong data in his calculations.

The Mathematician

The math teacher was Johannes Kepler. Like Copernicus, Kepler had been educated for a career in the Church. However, Kepler considered God not as a divine being, but as the creative power that ran the universe. He firmly believed that if he could understand the language of the universe, mathematics, he might discover its secrets.

To perform proper calculations, however, he needed more accurate observations. Such data was almost non-existent at the time, since most measuring instruments were not well built. Surely, somewhere there must be some good data he could use.

Kepler thought some more. There was one place: in Denmark, on the island of Hven. It was here that the famous astronomer and Danish nobleman Tycho Brahe had built his magnificent observatory to study the heavens.

Tycho's work had become legendary. He had spent many decades measuring precise positions of stars and planets. This was just what was needed, Kepler thought. Surely, Tycho would share his observations and help him unravel the secrets of the universe.

But Kepler lived a life of very meager means and could not afford to travel to Denmark to visit Tycho. Then in 1596, Kepler's luck began to change. That year, Tycho had a falling out with the Danish king and left Denmark altogether. After roaming about Europe for several years, Tycho settled in Prague, as the Imperial Mathematician in the court of the Holy Roman Emperor Rudolph II.

Finally, Kepler thought, Tycho was near enough to visit. He could go to him, get the necessary data he needed, and carry out the work of his dreams. So he set out to visit Tycho. It was January 1, 1600.

On To Prague

A month later, Kepler arrived in Prague. As his carriage turned into the Castle of Benatek where Tycho lived, Kepler gasped in amazement. What he saw was beyond his wildest imagination. Tycho lived in a palace!

Kepler stepped from the carriage. A thin, shabbily dressed man, he gazed in awe at the magnificent castle surrounding him. How could anyone be so lucky to live such a life of luxury?

And then, from among his crowd of assistants and relatives, Tycho appeared. He was a large, rotund man—bald, with a large curled moustache, and impeccably dressed. As he turned and walked toward Kepler, the sunlight glistened off his nose. It was made of gold and silver alloys to replace his real nose, which he had lost in a duel with a classmate over who was the better mathematician.

Giraudon/Art Resource

Giraudon/Art Resource

Instruments such as this ivory tablet sundial* (above) *were used to tell time in the sixteenth century. As the sun moved daily across the sky, its light cast a shadow of the* gnomon *(the thin filament over the center of the sundial) in different directions. Far left: Tycho Brahe (1546–1601), the Danish nobleman and astronomer, was known far and wide as one of the greatest astronomical instrument makers in the world. His tools helped him measure the positions and motions of the stars and planets. Left: Johannes Kepler (1571–1630) was the mathematician who first learned the correct paths the planets take on their way around the sun, and the mathematical laws that govern them. At the age of thirty-seven he wrote* Somnium (Dream), *perhaps the first true science-fiction story in history.

Kepler was welcomed, and was given a tour of the famous observatory. As he walked beside Tycho, Kepler grew more and more excited, for he knew that before him were instruments that gathered data ten times more accurately than those used by Copernicus.

Where's the Data?

After the tour, Kepler returned to his room with a smile. His decision to come to Prague had been a stroke of genius, and he knew it wouldn't be long before he would see Tycho's data and could begin to create a new model of the solar system.

But Tycho, it seemed, had other ideas. At mealtimes, conversation would turn to astronomical observations, and Kepler would sit up attentively. Then, just as quickly, Tycho would lose interest and go on to other, less important, topics.

Day after day, week after week, it continued. To Kepler it was sheer torture, and he became more and more frustrated and angry. Then, just as he had lost all patience and had packed his bags, Tycho offered him a job.

His assignment was to use Tycho's data to calculate an orbit for Mars, a planet whose convoluted motions had puzzled stargazers for centuries. Kepler figured he could solve the problem in no time, so he began his work.

In this engraving, Tycho Brahe sits with his great mural quadrant and other instruments he used at his observatory at Uraniborg. Tycho is pointing to a hole in the wall through which he observed the stars. Below are two clocks and his alchemical laboratory.

The Bettmann Archive

Snark/Art Resource

Planisphere of the universe, according to Brahe; it retains the Earth at the center. The moon and sun orbit the Earth while the planets orbit the sun at various distances—a complicated, but not unreasonable, scheme.

After one of Tycho's typical feasts on October 13, 1601, the great astronomer became terribly ill. Day after day, Kepler stopped in to talk, but the astronomer's condition continually worsened. Somehow, Tycho knew the end was near, and he was heard to mumble over and over again: "Let me not seem to have lived in vain.... Let me not seem to have lived in vain...."

Then, eleven days later, the end came; Tycho was dead. Within weeks, Kepler was offered the job as Tycho's replacement. Now all the cherished and protected data of the great astronomer would finally be his.

But Kepler had other things on his mind: He had become obsessed with seeking the mathematical relationships of the universe. Now Mars was taking up every waking hour of his time. Devising an orbit for this planet wasn't as easy as he had expected. Day after day, he pored over the data, drawing circle upon circle upon circle in hopes that just one—any one—might fit.

The Discovery

Months passed into years, and all he had to show for his work was nine hundred pages of calculations and seventy worthless orbits. And then, around Easter of 1605, Kepler decided he had seen enough circles for one lifetime. He concluded that all he had left to try was "a single cartful of dung"—a stretched out circle called an ellipse.

In absolute frustration, Kepler began to draw his ellipse over the data. As he did, his eyes lit up—it fit beautifully! No wonder people had been confused. Mars didn't orbit the sun in a circle at all. It orbited in an ellipse!

In a single moment of unrivalled genius, Johannes Kepler solved a problem that had confounded astronomers for centuries. In unbridled joy, he sketched on his work the goddess of victory riding her chariot above the clouds. "The truth of nature, which I had rejected and chased away," he later recalled,

The Bettmann Archive

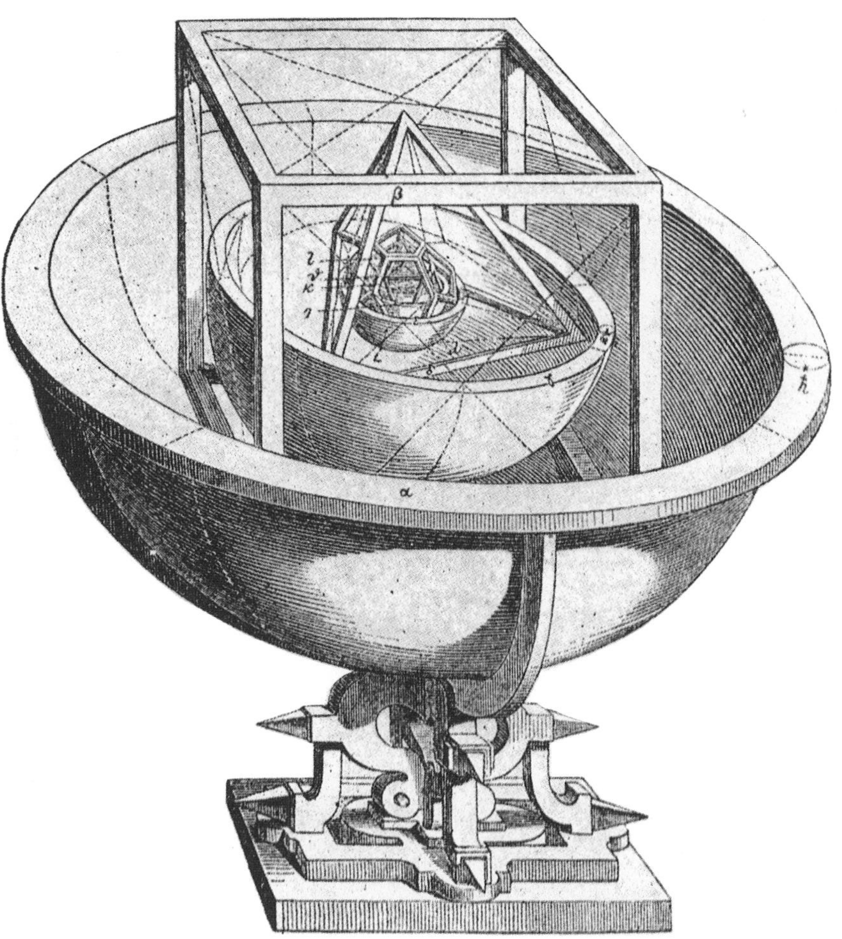

"returned by stealth through the back door, disguising itself to be accepted . . . Ah, what a foolish bird I have been!"

THREE LAWS

Armed with the knowledge that Mars orbited the sun in an ellipse rather than a circle, Kepler set out to learn the motions of the other planets. Over the years, he discovered three fundamental mathematical principles that explain how our planetary neighbors operate. Today, they are known as Kepler's Laws of Planetary Motion.

The first law states that the sun lies not at the center of a planet's orbit, but is offset slightly at the "focus" of the ellipse. The second states that planets move faster when closer to the sun, and slower when farther away. And the third states that the inner planets move faster on their orbits than the outer planets.

Kepler couldn't imagine why these fundamental planetary laws should exist at all. He wondered how the planets knew where they were at any one time, and how they knew when to speed up and slow down. Was there some invisible force that came from the sun and controlled the motion of the planets?

Kepler had no idea. But it wouldn't be long before someone would discover an answer.

Far left: ***German astronomer Kepler discussing his discoveries of planetary motion with his sponsor, Holy Roman Emperor, Rudolph II. Though the Emperor was interested mainly in mysticism and astrology, he also encouraged the pursuit of the science of astronomy.*** **Above:** ***In his 1619 publication of*** **Harmonices Mundi (Harmony of the World),** ***Kepler put forth his early ideas of the structure of the planetary system. It featured five regular solids between the spheres of the various planets.***

chapter four

THE EARTH MOVES

While in Florence, Galileo Galilei (1564–1642) lived and worked beneath this terrace at the Tower of the Rooster.

Alinari Archive/Art Resource

The year was 1609. A cool breeze was blowing on this May evening as a carriage rattled briskly through the cobblestone streets of Padua, Italy. In the carriage was Galileo Galilei, returning home from Venice with news of a remarkable invention.

While in Venice, Galileo had received word that a Dutch optician had built an eyeglass that could make distant objects appear nearby. If this were true, Galileo thought, then a building a mile away could be seen as clearly as if it were right next door. When he returned home, he sketched out plans to build one for himself, and then turned in for the evening.

Early the next morning, golden sunlight streamed through the windows of Galileo's laboratory. Soon, the door swung open, and in walked the scientist eager to begin his work.

Galileo was a plump, bearded man, a man confident in his

Alinari Archive/Art Resource

ability to solve the mysteries of nature. During his lifetime he had studied such varied subjects as medicine, mathematics, and mechanics, and he had built numerous instruments in many different fields, including compasses, surveying tools, and an instrument for taking a person's pulse.

It was his knowledge of optics that he relied on now. Unwrapping an assortment of eyeglass lenses he had brought from Venice, Galileo excitedly began his work. He knew that the Dutchman's glass used two lenses, one convex and one concave, and he began assembling and testing various pieces.

Above: ***In this room, Galileo first developed a telescope to view the moon.*** **Inset:** ***This military compass was made by Marcantonio Mazzolen, hired by Galileo to make his instruments while living and working in Padua.***

The New Telescope

Soon, he had fit a lead tube with lenses and was ready to try it out. Standing at the window, he raised the glass to his eye and aimed it at a distant building. What he saw was remarkable. The building appeared three times closer than it actually was!

This was an amazing invention. No wonder people had used it for years to spot ships long before they reached port. But Galileo realized that this tool was far too valuable to use only for such limited purposes. He dreamed of building telescopes of even greater magnifying power—eight, fifteen, twenty times—and to use them not for looking at the ground, but for studying the sky.

It wasn't long before Galileo had a telescope far better than any in existence. Resting in a cradle on a stand, Galileo's new instrument could bring the heavens thirty times closer. What discoveries lay in store he could hardly imagine.

As darkness fell that evening, Galileo aimed the tube toward the thin crescent moon sinking low in the west. His heart beat in anticipation of what he might see. When he looked through the tube, he gasped. Rubbing his eyes, he looked again. He couldn't believe what he was seeing.

There it was—the moon. But it hardly resembled the descriptions handed down by the great thinkers of ages past. The moon wasn't the smooth, perfect sphere many thought it to be. It was pock-marked with holes, and its surface was cracked and splintered.

Not only that, but there were mountains as well. These weren't just rolling hills either. Galileo measured their shadows and calculated that they must be as tall as four miles (six kilometers) high.

On this momentous evening, Galileo had found the moon to be a world similar to the Earth. But how similar, he wondered. Did it have water and air? Did it have people and trees and animals, or was it a silent, dead world? The questions and ideas raced madly through his mind as he peered through the tube at his new discovery.

It wasn't long before the moon set in the west, and Galileo aimed his new telescope upward toward the stars. What he saw was astounding: thousands and millions of stars scattered about the sky, stars never before seen by human eyes. And every one of these was but a pinpoint of light. If this telescope magnified their images thirty times and still

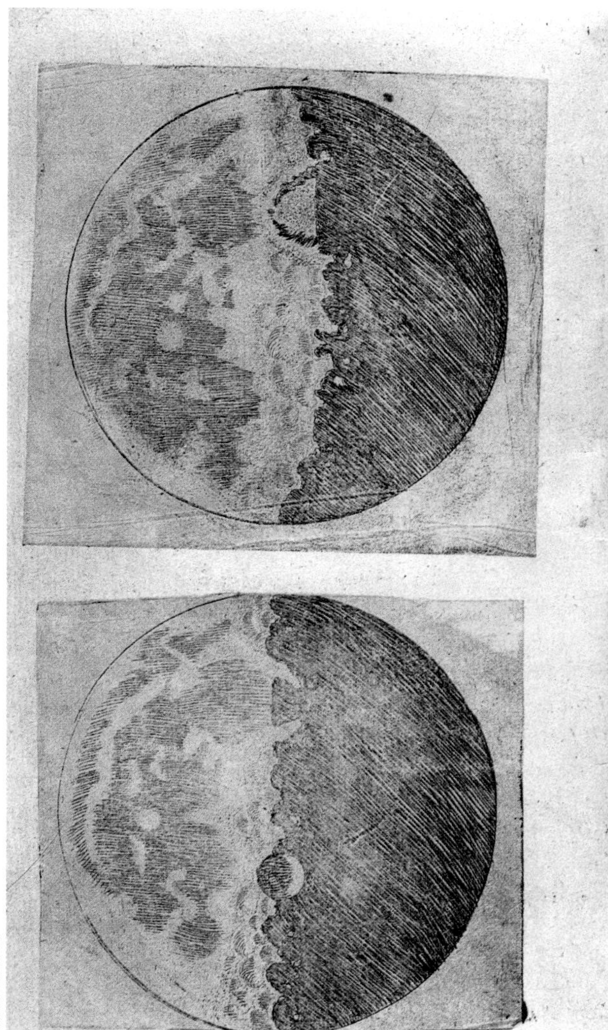

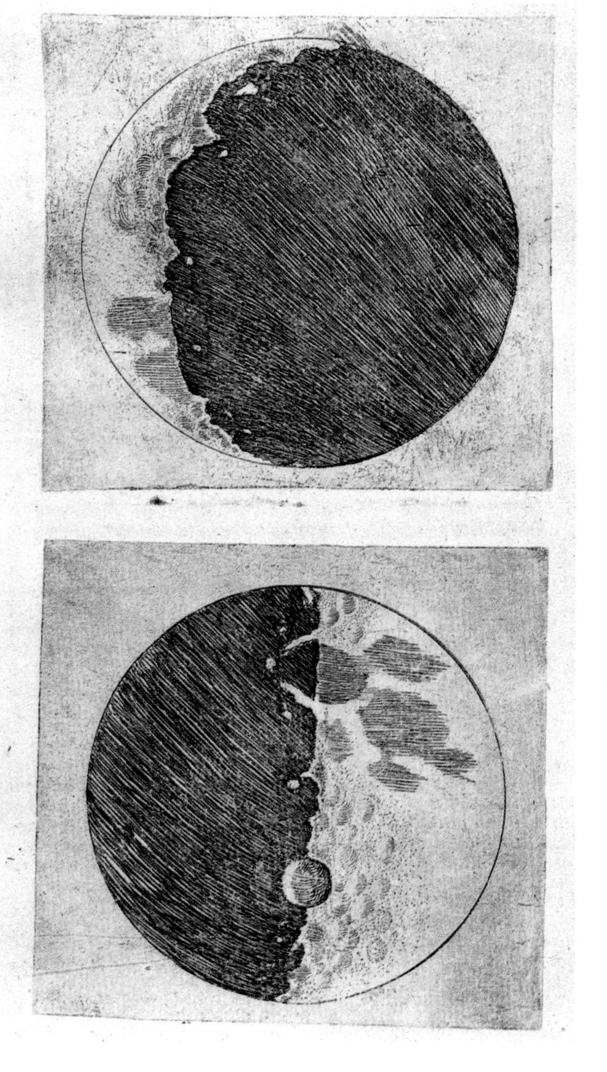

Above: ***A portrait of Galileo Galilei. While Galileo did not invent the telescope, nor was he the first to turn it toward the heavens, he made discoveries that revolutionized the science of astronomy.*** **Left:** ***These lunar maps, sketched in Galileo's own hand, show the remarkable detail seen through his tiny telescope.***

This astrolabe was made in Italy in the late sixteenth century. While Galileo did not design it, some believe he may have used it to measure celestial positions while in Padua from 1592 to 1610. Opposite page: Two of Galileo's original telescopes and the broken lens through which he discovered the four moons of Jupiter are all preserved in the Tribuna de Galileo in Florence.

they looked like points, he thought, the stars must be tremendously far away.

What a fantastic new tool this was. Night after night, Galileo probed the mysteries of the universe from his garden and wrote and sketched his many discoveries in a journal.

WORLDS OF JUPITER

On January 7, 1610, Galileo noticed the bright planet Jupiter in the sky, and he aimed his telescope in its direction. By now, he knew that each new object he observed held remarkable surprises. When he saw Jupiter, he wasn't disappointed.

Through his tube, the planet appeared quite different than the stars; it was a small round disk. And nearby, Galileo noticed three bright points of light—two to its east and one to its west. He made a sketch of the planet and these stars and continued observing.

The next night he looked at Jupiter again, but something was different. The three tiny stars near Jupiter were not where they had been the night before. Tonight all three were to the west of the planet. If Jupiter had moved past them as it orbited the sun, he reasoned, then it didn't really move as astronomers had predicted. He was anxious to see what would happen next.

The following night was cloudy, but by the tenth of the month, the skies had cleared. When he observed Jupiter that night, he couldn't believe his eyes. There were only two stars visible, and both were to the east of the planet. Galileo checked his sketches and looked once again. Was he going crazy? Was Jupiter somehow swinging back and forth like a pendulum in front of these three stars? Or was something totally unknown going on?

Night after night, he watched Jupiter and the three little stars, never knowing quite what to expect next. Then on the thirteenth, he aimed the telescope toward Jupiter once again. But that night there weren't two stars; there weren't three; there were *four*, one in the east and three in the west.

Galileo had suspected it all along, but finally he knew for sure. These intriguing points of light weren't stars at all, they were moons that orbited their parent planet.

To Galileo, this was absolute proof that Copernicus was right. The Earth was not the center of all things. It couldn't be, for here were four objects orbiting another world. Jupiter it seemed, was a miniature model of our entire solar system.

Two months later, Galileo published his amazing discoveries in a short book he titled *Sidereus Nuncius (The Celestial Messenger)*. His observations caused quite a stir around Europe, and were even discussed as far away as China. But Galileo didn't stop there. He soon turned his attention toward Venus.

EVEN MORE DISCOVERIES

He found that Venus appeared quite different than Jupiter. Not only didn't it have any moons orbiting it, but it seemed to change its size and shape as it moved about the sky—much like our own moon. At times, the planet appeared to be a large, thin crescent. Then, over several weeks, it would shrink in size and display a "quarter" phase. And, just before disappearing into the glare of the sun weeks later, it would appear as a tiny, full disk. Galileo concluded that this could only be explained if Venus moved around the sun and not around us.

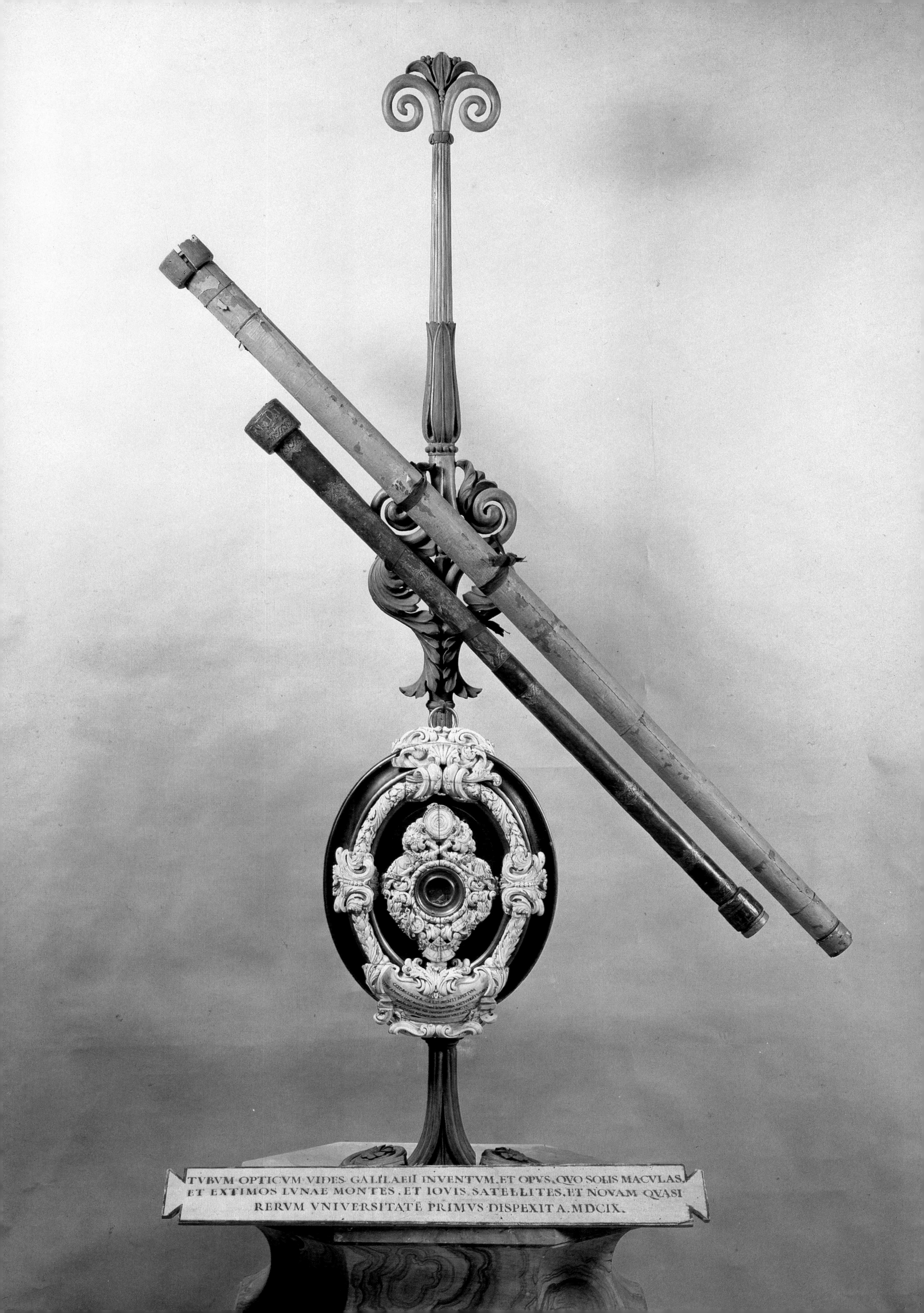
TVBVM OPTICVM VIDES GALILAEII INVENTVM, ET OPVS, QVO SOLIS MACVLAS,
ET EXTIMOS LVNAE MONTES, ET IOVIS SATELLITES, ET NOVAM QVASI
RERVM VNIVERSITATĒ PRIMVS DISPEXIT A. MDCIX.

While telescopes had been used for years as navigation devices, Galileo was the first man to point one toward the heavens. What he found revolutionized the study of astronomy.

Next, Galileo found that the sun was not the unblemished fire in the heavens as was dictated by the wisdom of the ages. It had dark spots appearing against its bright face, and these moved about from day to day. Even the sun itself was rotating!

Galileo knew that these discoveries contradicted the wisdom of the ancient philosophers, not to mention the teachings of the Church. Galileo was a good Catholic and wished no harm to come to his Church or to its teachings. But he was also a first-rate scientist, and he knew the facts when he saw them: The Earth and planets orbited the central and motionless sun; not the other way around.

The Trouble Begins

Within months, Galileo moved to Florence to accept a new position. At the same time his book had become quite popular, drawing a great deal of attention to his radical ideas. The Pope, however, had become angry at Galileo's apparent disregard of Church policy; so to make peace with his friends in the Church, Galileo went to Rome.

His trip was disastrous. On February 25, 1616, he was ordered to stop defending the "false" doctrine of a spinning Earth orbiting a motionless sun. In June of that year, he returned to Florence to continue his research and wait for a new opportunity to state his case once again.

Galileo's patience lasted only a few years. In 1624, he decided to write a book describing the two prevailing "world systems" without disobeying the strict Papal decree. He called it *Dialogo Dei Due Massimi Sistemi (Dialogue Concerning the Two Chief World Systems).*

The book was designed as a friendly conversation about the theories of Copernicus and Ptolemy. The discussion was dominated by a character named Salviati who eloquently defended the Copernican theory. A second character, Sagredo, was an intelligent, but uninformed believer in Ptolemy's system, but was willing to listen to both sides of the argument. A perfectly innocent idea, Galileo thought.

Where he went wrong was in using a third character who staunchly defended Aristotle and Ptolemy, advanced the Pope's favorite arguments against Copernicus, and always had the last word. This poor, bumbling fool was named Simplicio.

Bridgeman/Art Resource

Above: ***In 1633, Galileo was tried, convicted, and sentenced by the Catholic Church for spreading "false doctrines" about the nature of the world and the universe. It wasn't until the 1980s that the Church reconsidered its case against Galileo, and absolved him of all crimes.*** **Opposite page:** ***Designed by Giulio Goddini, this marble tomb at the Church of Santa Croce in Florence now marks the burial site of Galileo.***

The Trouble Continues

When the *Dialogue* was published in 1632, Church officials were furious. Not only did the book ridicule the Pope, but it did so in Italian rather than Latin, and could be much more widely read. When the Church ordered that all unsold copies be sent to Rome, the printer replied that there were none left to send there.

Now, the Church had all the evidence it needed to bring charges against Galileo and force him to stop spreading "false doctrines." Galileo was in serious trouble.

In September 1632, Galileo was ordered to stand trial for heresy. By now, the last thing he wanted was to return to Rome. After all, it had been only three decades since Giordano Bruno was burned at the stake for similar conflicts with the Holy Office. But Galileo wisely decided against challenging the Church further and agreed to the trial.

In April of the following year, the interrogations began. Throughout the ordeal, Galileo must have secretly hoped that Church officials would reconsider their antiquated beliefs once they heard his scientific reasoning.

But such was not to be, for as Galileo soon learned, science and reason played absolutely no role in their decisions. When the trial ended on June 20, 1633, Galileo knew that he had been beaten.

The Sentence

Two days later he was led into the great hall of the convent of Santa Maria Sopra Minerva where he was to hear his fate. His spirit now broken, Galileo was forced to kneel before the court and retract his belief that it was the sun that stood still, and the Earth which moved around it. Immediately following, he was placed under house arrest, where he remained for the rest of his life.

In 1642, this once great and enthusiastic scientist died a lonely, tired old man in his home at Arcetri; he now lies buried in a marble tomb at the Church of Santa Croce in Florence.

GALILAEVS GALILEIVS PATRIC. FLOR.
GEOMETRIAE ASTRONOMIAE PHILOSOPHIAE MAXIMVS RESTITVTOR
NVLLI AETATIS SVAE COMPARANDVS
HIC BENE QVIESCAT
VIX. A. LXXVIII. OBIIT. A. CIↃ. IↃ. C. XXXXI.
CVRANTIBVS AETERNVM PATRIAE DECVS.
X. VIRIS PATRICIIS SACRAE HVIVS AEDIS PRAEFECTIS
MONIMENTVM A VINCENTIO VIVIANIO MAGISTRI CINERI SIBIQVE SIMVL
TESTAMENTO F. I.
HERES IO. BAPT. CLEMENS NELLIVS IO. BAPT. SENATORIS F.
LVBENTI ANIMO ABSOLVIT.
AN. CIↃ. IↃ. CCXXXVII

Legend tells that, as Galileo walked slowly and sadly from the courtroom on that fateful June day, he was heard to mutter under his breath five words that would echo defiantly for generations to come: "... and still the Earth moves."

A Budding Scientist

Galileo's death marked the end of an era in which new discoveries overthrew the beliefs of the ancients and aroused emotions in all. But it was also the beginning of an era, for in that very year a child was born in the Lincolnshire village of Woolsthorpe, England, a child who ultimately completed the astronomical revolution that had begun with Copernicus only a century earlier. His name was Isaac Newton.

That Isaac would one day grow up to become a great scientist must have surprised no one. As a child, he was impatient and moody and enjoyed spending time alone designing practical devices. During his young life, he built such things as a water meter and a sundial, and he invented a moving vehicle operated by only a pedal.

These were admirable qualities for a budding scientist, but they were useless for carrying out important tasks of the day. Being the

Above: ***Isaac Newton (1642–1727) has been called the most brilliant scientist in history. Not only did he formulate the universal law of gravitation, he also invented the mathematics known as calculus, and the reflecting telescope. He was knighted by Queen Anne in 1705.*** **Right:** ***Woolsthorpe Manor, the house where Isaac Newton was born. It was here during the time of the great Black Plague of London that Newton began his work on the theory of gravitation.***

Probl I

Investiganda est curva linea ADB in qua grave a dato quovis puncto A ad datum quodvis punctum B vi gravitatis suae citissime descendet

Solutio.

A dato puncto A ducatur recta infinita APCZ horizonti parallela et super eadem recta describatur tum Cyclois quaecunque AQP rectae AB (ductae et si opus est productae) occurrens in puncto Q, tum Cyclois alia ADC cujus basis et altitudo sit ad prioris basem et altitudinem respective ut AB ad AQ. Et haec Cyclois novissima transibit per punctum B et erit Curva illa linea in qua grave a puncto A ad punctum B vi gravitatis suae citissime perveniet. Q. E. I.

The Bettmann Archive

Left: ***The solution to the problem of* brachystochrone, *or the curve of quickest descent, by Newton.*** **Below:** ***While sitting under a tree in 1666, Newton watched an apple fall. He began to wonder about the force that attracted the apple downward, and the force that held the moon, the Earth, and the planets in their orbits. From this observation, he developed the law of gravitation.***

eldest son, Isaac was expected to help run the family farm, and that included tending sheep.

But Isaac was not a shepherd. Every day, the flock would wander away from him while he sat totally lost in thought. And every evening, someone would have to go out to the field to round up the sheep for him. One evening, the sheep were all accounted for, but it was Isaac who had wandered off!

It was his aptitude for science and mathematics that took him to Trinity College at Cambridge University in 1661, but it was his concentration and absent-mindedness that soon became legendary. Isaac often became so engrossed in his studies that he wouldn't hear the dinner bell ring, and his classmates would have to drag him off to eat. Some would even wager if he'd show up at all.

Then, in 1665, the Black Plague struck in London and Isaac was sent back home to avoid the disease.

Newton's Apple

One afternoon, while his mother was preparing the evening meal and his brothers and sisters were helping to harvest the crops, Isaac sat beneath a tree outside his house. As usual he was deep in thought.

He was wondering why the planets continued to orbit the sun rather than flying off in a straight line. He recalled as a child having whirled around him a stone tied to a cord. As long as he held the cord, the stone continued to go in circles, but when he let go, the stone always flew off in a straight line away from him. What was the "cord" that held the planets in their orbits around the sun?

The Bettmann Archive

Just then, a gust of wind blew an apple from the tree. When it hit the ground with a thud, Isaac's mind made a connection: Why did the apple fall where it did? If the wind were blowing harder, would the apple have fallen farther from the tree? Why not even hundreds or thousands of miles away? And what if the wind blew so hard that the apple never actually hit the ground, but instead kept on going around the Earth?

Most people might have dismissed these questions as mere curiosities, but not Isaac. He realized that something, some mysterious force, was coming from the center of the Earth and was keeping the apple from flying off into space. That same force, he reasoned, must also be what keeps the moon and planets in their orbits.

The Bettmann Archive

Newton bought his first prism at a country market, and was forever mesmerized by its effects. He used it to study how white sunlight is dispersed into a wide band of colors.

During the next few years, Newton went on to examine Kepler's Laws of Planetary Motion for clues as to why the planets orbited the sun in ellipses, and why those nearest the sun moved fastest, while those farther away moved more slowly.

He concluded that space is pervaded by an invisible force, which Newton called gravity—just the "cord" he was looking for. Gravity, Newton suggested, seems to originate in the center of all bodies, and tends to pull all things inward. He discovered that the more mass an object had, the greater its gravitational pull would be. And, he found that the farther apart two objects were, the less force either one would feel from the other.

While he checked and rechecked his theories, he began working on other projects. One of these was the telescope. Newton was fascinated by telescopes, but was terribly bothered by color distortions he could see through the lenses of the day.

To figure out what was going on, he passed a shaft of white sunlight through a triangular piece of glass called a prism. Prisms had been known to create a beautiful beam of colors since the days of Aristotle, but Newton was curious why the colors came out the other end not in a circle, but in an oblong form.

Newton continued experimenting to learn if perhaps the glass somehow added these colors to the white light. When he held another prism upside down in the beam of colored light, something equally remarkable happened. Now, the colors were turned back into a shaft of white light.

What Newton realized was that the colors had been there all along, but that when mixed together they appeared white. The glass bent the colors differently and spread them out into a spectrum. This must be the same effect that was caused by the lenses of a telescope.

The only solution he could imagine was a telescope without lenses. What then would gather starlight and create an image? Why not a mirror?

It was this idea with which Newton began to experiment. He began by making a telescope with a one-inch (2.5-centimeter) diameter, concave mirror at the bottom end. The mirror gathered light from the sky and sent it back up the tube. But he couldn't put his eye there to see it because his head would block the incoming starlight. Instead he put a second, flat mirror at a forty-five degree angle near the top to send the light out the side of the tube, where he placed an eyepiece to study the image.

While Newton never made any systematic observations through his instruments, he did produce the first "reflector" telescope in history. It could magnify images thirty-five times, and showed scientists some marvelous views of the sky that were completely free from the color distortions of even the best "refractor" telescopes of the day.

A Unified Theory

Meanwhile, Newton continued working on his theory of gravitation. He found that he was mathematically able to explain the observations of Tycho and the laws of Kepler, and could view the entire planetary family as a closed system governed by very clear and simple laws.

Not since Aristotle had anyone been able to synthesize all that was known about planetary motion into one grand, unified theory. Newton, however, didn't feel his theory was important enough and didn't bother to publish his work.

Fortunately, astronomer Edmund Halley (whose name is attached to the now-famous comet) finally convinced Newton to write a book and share his theories with the world. He even offered to pay for its publication out of his own pocket.

The book Newton wrote was called *Principia Mathematica Philosophiae Naturalis (Mathematical Principles of Natural Philosophy)*. Now known as simply *Principia*, it took fifteen months for Newton to compile, and has been described as the greatest mental effort ever made by one man.

Whether Newton would have agreed with this statement is debatable, for until he took his last breath, Newton maintained that his contribution amounted to just one tiny piece of the cosmic puzzle. Just before his death in 1727, he wrote these words that will live forever:

"I do not know what I may appear to the world; but to myself I seem to have been only like a boy, playing on the seashore, and diverting myself, in now and then finding a smoother pebble or a prettier shell than ordinary, while the great ocean of truth lay all undiscovered before me."

Through the new laws of gravitation and optics discovered by this scientific giant, that great ocean was about to be probed for the first time.

Isaac Newton's first reflecting telescope, which he gave to the Royal Astronomical Society. Below it is the extra mirror he ground himself.

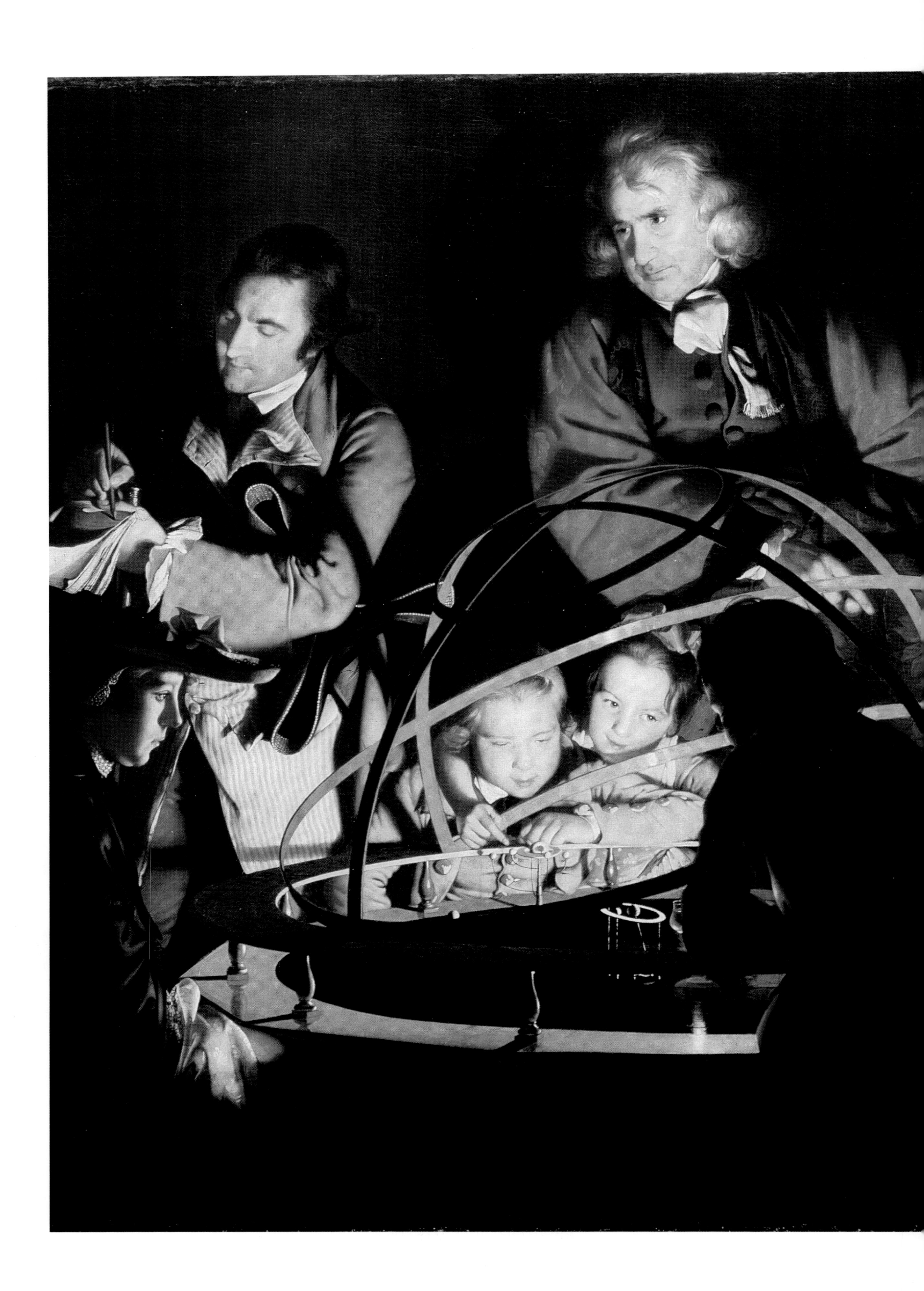

chapter five

THE STAR HUNTERS

It was a cold, clear December evening in 1779. The stars sparkled brilliantly in the sky over Bath, England. From a telescope set up on the road outside his house, Friedrich Wilhelm Herschel was observing the mountains on the moon.

Just then, a well-dressed passerby stopped and watched Herschel. The stranger asked if he might take a glimpse through the telescope.

"Why, of course," Herschel replied, and stepped aside to let him see. When the stranger peered through the telescope, he gasped in amazement. He had viewed the moon through telescopes before, but never had he seen anything quite so remarkable.

After they chatted for awhile, the man thanked Herschel, wished him a good night, and went on his way. Herschel continued observing the moon.

Many orreries up to three feet (ninety-one centimeters) across, showing the motions of six known planets and their satellites, were made in England during the eighteenth century. Their name is usually attributed to John Rowley who, in 1712, made such a device for Charles Boyle, 4th Earl of Cork and Orrery.

Bridgeman/Art Resource

The next morning, the stranger appeared again at Herschel's door. His name was Dr. William Watson, a Fellow of the Royal Society and a member of the Bath Literary and Philosophical Society. He had been so impressed with Herschel's interest in astronomy and his telescope-making skills that he invited him to join the Bath Society.

Herschel was quite pleased. For years he had been a successful music teacher and conductor, and most people considered his passion for astronomy just a strange quirk. Finally, here was someone who took his astronomical interests seriously.

So Herschel joined the Society, and over the next year, submitted more than two dozen papers about his research. And he continued to construct larger and larger telescopes in his spare time.

Soon, Herschel had amassed the finest collection of telescopes in the world. His largest was twenty feet (six meters) long and supported a mirror twelve inches (thirty centimeters) across. With these instruments, he had catalogued nearly all the stars visible to the unaided eye and had begun a second listing of even fainter stars. But it was on a cool spring evening in 1781 that Herschel's name would go down in astronomical history.

Above: ***Sir William Herschel (1738–1822) began his career as a conductor, composer, and musician, but quickly became interested in optics and astronomy. During his lifetime, Herschel built the largest and most productive telescopes in the world.*** **Right:** ***The giant forty-eight-inch (1.2-meter) diameter reflector built by Herschel in 1789. Unfortunately, the telescope was much too big for its mounting to be successful. He built it with a grant from King George III, who enjoyed taking guests to look at the instrument.***

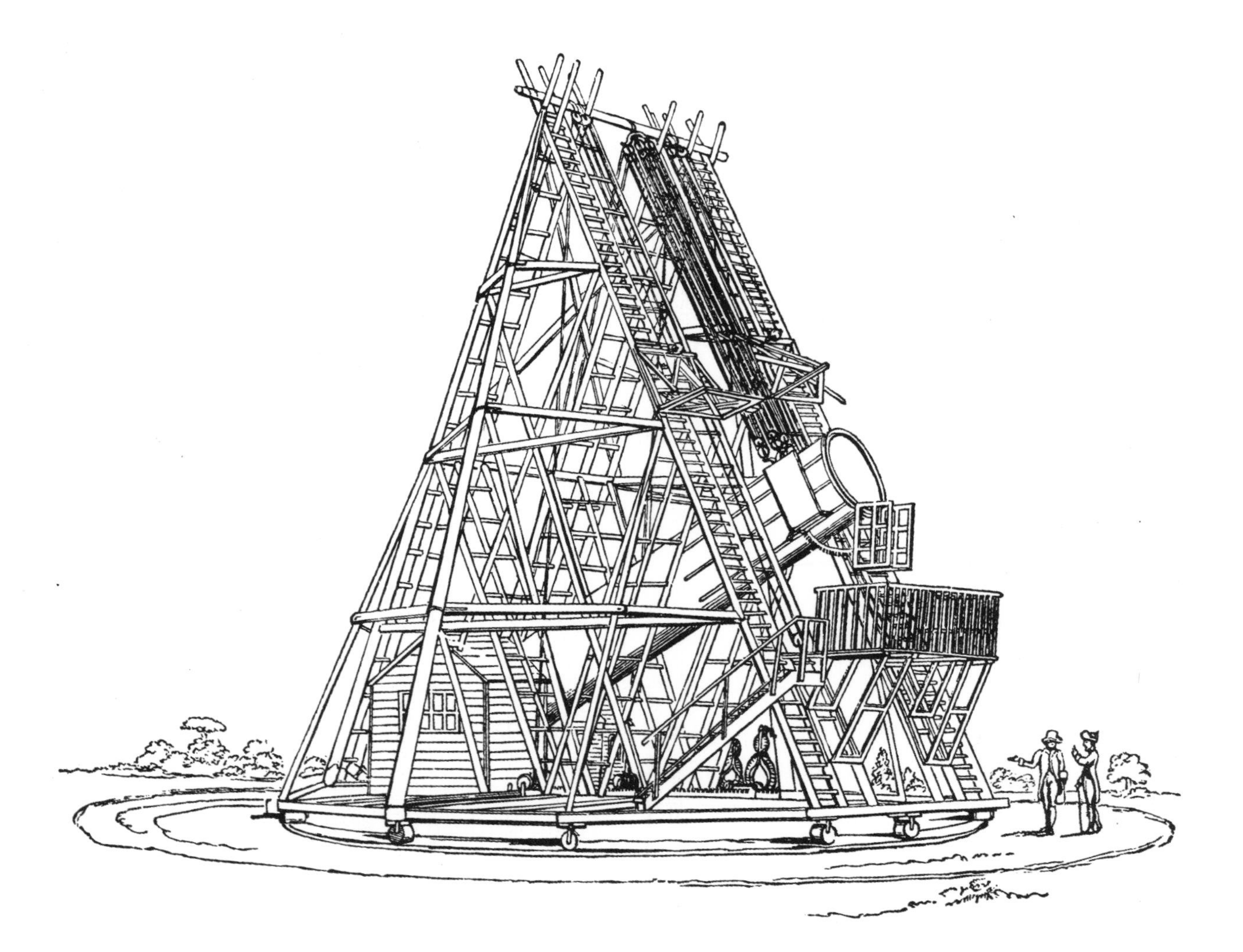

HERSCHEL'S "COMET"

On the evening of March 13, Herschel swept his giant seven-foot (two-meter) long reflector telescope slowly over the stars, noting each in its place. And then, between 10:00 and 11:00 P.M. Herschel abruptly stopped the telescope.

Near one of the stars in the constellation Gemini, something peculiar appeared. Herschel switched to a higher power eyepiece to study the object more carefully. He found that the object appeared larger than the stars, suggesting that it was much nearer.

Herschel thought that perhaps it was a comet, since comets sometimes appear unexpectedly and drift slowly among the stars. To see if it would move during the night, Herschel measured its position carefully and went about his other work.

Then, just as dawn was beginning to break, Herschel aimed his telescope back toward the object and measured its position once again. Just as he expected, it had moved slightly among the stars. Herschel sent word of his comet discovery to Oxford Observatory and to Dr. Watson, and didn't think much about it.

But other astronomers weren't so complacent about the discovery; for Herschel's "comet" was unlike any they had ever seen before. It wasn't fuzzy, and it didn't have a tail. In fact, they determined that it wasn't a comet at all, but a new planet!

Suddenly, the solar system became twice as large as anyone had ever thought. The new world circled the sun far beyond the orbit of Saturn, taking eighty-four years to go once around. Astronomers named it Uranus.

THE MOST FAMOUS ASTRONOMER IN THE WORLD

For his remarkable discovery, Herschel was presented with the Copley Medal, the highest honor given by the Royal Society, and was elected a Fellow of Society. In addition, King George III paid Herschel a handsome salary to become his "personal" astronomer—under the condition that the Royal Family be allowed to peer through his telescopes once in a while.

What more could Herschel want? He was now recognized as a world-class scientist and could devote all his time to the study of the heavens.

As he continued his study of the universe, Herschel grew intrigued with the many nebulous objects scattered among the stars. Through his telescopes, Herschel found that some were indeed wispy clouds of material, while others were groups of individual stars.

Herschel then began to wonder whether it was possible that these "nebulae" were actually island universes, and that our sun was located in a similar system called the Milky Way? And if it was, what was its shape? And how many stars made it up?

THE STAR COUNTER

Herschel had many questions, but very few answers. So, he decided to count stars around the sky to determine the shape and size of the universe. He reasoned that the more stars he saw in any one direction, the farther he must be looking.

During his long and tedious star counts, Herschel was often assisted by his sister, Caroline. She eventually began her own astronomical research, and discovered eight comets. In 1828 she was awarded the Gold Medal of the Royal Astronomical Society for her contributions to the science.

North Wind Picture Archive

Right: ***By 1750, the study of astronomy was a favorite pastime of the upper classes.*** **Below:** ***Using a telescope made by Joseph von Fraunhofer, Friedrich Wilhelm Bessel was the first to accurately calculate the distances between stars and Earth.***

Bridgeman/Art Resource

Night after night, month after month, he counted the stars. In some places he saw only one or two. In others, particularly along the hazy band of the Milky Way, he found hundreds or thousands. In 1785, he put his discoveries together into one simple picture of the universe.

Herschel described our universe as an elongated, disk-shaped system of millions of stars. The hazy band of light known as the Milky Way was actually the blended light of millions of stars forming a disk. Because this hazy band seemed nearly the same brightness no matter where he looked, our sun must lie near the disk's center. And elsewhere lie hundreds of other similar systems, which appeared to him as wispy patches of light around the sky.

From these time-consuming, painstaking star counts, Herschel had determined the structure of the entire universe. But its size had him puzzled, for there was no way of measuring distances among the stars. Astronomers would have to wait nearly half a century to learn the distances to the stars.

THE COSMIC YARDSTICK

Ironically, the principle of measuring stellar distance was known to the Greeks thousands of years earlier. They knew that a nearby star should appear to shift its position slightly, since we on Earth view it from different directions as our planet orbits the sun each year. The same effect occurs when we hold up a finger and look at its position against the distant background first with one eye, and then the other.

Bildarchiv Foto Marburg/Art Resource

Nobody had ever detected such a stellar "parallax," which meant that the stars must be tremendously far away. But how far no one could say. A remarkably accurate telescope was needed to determine this.

An astronomer in East Prussia named Friedrich Wilhelm Bessel owned just such a telescope. It was a magnificent optical instrument with a ten-foot (three-meter) long polished mahogony tube, made by the famous Austrian scientist, Joseph von Fraunhofer. If the parallaxes of stars could be measured at all, Bessel thought, *this* was the telescope to do it.

But Bessel was a perfectionist, and he knew the instrument could be improved even more. "Every telescope is made twice," he wrote, "first in the workshop, and then by the astronomer." And this telescope, as wonderful as it was, was no exception.

So Bessel made calculations to correct for any misalignments of the telescope lenses, the motions of the Earth, and distortions caused by our turbulent atmosphere. Finally, in August 1837, Bessel was ready to begin his work.

He chose to study a faint star named 61 Cygni. 61 Cygni was known as the "flying star" because, since the days of Kepler it had changed its position by nearly the width of the full moon—far more than any other star in the sky. It was this rapid motion which indicated to Bessel that the star must be nearby.

Night after night, he sat in the observatory and very carefully measured the star's position against two background stars. By day, he shut himself away in his study and worked on figures to eliminate the tiny errors which might somehow influence his measurements.

Within months, Bessel knew he had succeeded; yet he continued working to get an entire year's worth of data. In December 1838, he announced his results to the world. 61 Cygni was more than sixty trillion miles (ninety-six trillion kilometers) away—so far, that its light travelling at 186,282 miles (300,600 kilometers) per second took more than ten years to reach Earth!

Astronomers were astounded. They had known the stars were distant, but no one had ever imagined they were that far away. What was even more amazing was that 61 Cygni was one of the nearest stars to us.

THE MYSTERIOUS DARK LINES

Such fantastic discoveries were common during this time. Scientists had found an entirely new planet, measured the distances to the stars, and determined the shape and structure of the universe. In 1844, the French philosopher Auguste Comte began to take stock of the limits of human knowledge.

One of the things Comte believed must forever remain unknowable was the chemical makeup of the stars. He reasoned that it was impossible to gather a sample of a star and return it to Earth for laboratory analysis. But what Comte didn't know was that thirty years earlier, a discovery had been made that would ultimately prove him wrong. The discoverer's name was Joseph von Fraunhofer.

Joseph von Fraunhofer (1787–1887) was a German optician and physicist whose instrument-building skills were known far and wide. He built the instrument with which Bessel measured the first stellar parallax.

When Fraunhofer studied the sun's spectrum under high magnification, he found that it did not appear as a smooth band of color, as he had expected. It was, instead, split by hundreds of thin, dark lines.

What these lines meant remained a puzzle until the German physicist Gustav Robert Kirchoff became intrigued by them several decades later. In his laboratory at the University of Heidelberg, the bearded professor set up experiments to solve the puzzle. He used a spectroscope, an instrument that combines a prism with a small telescope to break white light into its colorful spectrum and to examine those colors. When Kirchoff heated different chemicals and viewed their light through the spectroscope, he found that none produced the smooth band of colors he had expected. Instead, they produced only thin, brightly colored lines. No two elements created the same pattern. Sodium produced two prominent lines; barium showed four; and iron displayed more than a thousand.

Next, he vaporized these elements and sent a shaft of white light through the gases. The same lines appeared, but now they appeared dark against the familiar colors of the rainbow! He thought that the gases must now be absorbing the same colors from the spectrum, and that the lines were like fingerprints that can identify the chemical in the gas!

Suddenly, Kirchoff understood why the sun's spectrum was crossed by dark lines. The sun was a giant chemical factory. Its atmosphere was made of cooler gases, which absorbed certain colors from the sun's spectrum.

It wasn't long before Kirchoff and his assistants began to pick apart the solar spectrum

Fraunhofer found dark lines in the spectrum of the sun. His research led to the development of an entire field of study known as astronomical spectroscopy.

The N.Y. Public Library

North Wind Picture Archive

Courtesy Celestron International

***Far left:* Gustav Robert Kirchoff (1824–1887) developed laws of spectroscopy that explained how spectra are created. His work led to an understanding of the chemical and physical makeup of the stars. Below: J. W. Draper (1811–1882) was a professor of chemistry and physiology at the University of New York. In 1840, he became the first person to create a photograph of the moon.**

and identify the chemical fingerprints of dozens of elements. On December 15, 1859, he announced his discovery to the Berlin Academy—a remarkably accurate way to probe the chemical makeup of the sun and stars, without leaving the safety of the laboratory here on Earth.

REPLACING THE EYE

Around the same time, astronomers were trying their hand at another amazing tool. It began one night in March 1840, as the American doctor J.W. Draper carefully aimed his telescope toward the moon. On this night, Draper had no plans to peer through the eyepiece at our cosmic neighbor. Instead, he placed a piece of glass coated with light-sensitive chemicals at its focus. For twenty minutes, he let moonlight fall onto the plate and then washed the coating away with a chemical bath. When he was finished, Draper had created the first photograph ever taken of a celestial body.

Photography was a truly remarkable invention, for it could permanently record whatever appeared in a telescope. No longer did astronomers have to sit for hours with their eyes rigidly fixed to the eyepiece. No longer did they need to rely on inaccurate sketches or biased interpretations. Now, a tool existed that could make their work more reliable and objective.

The Bettmann Archive

Right: ***The Tarantula Nebula, also known to astronomers as NGC 2020, is a cloud of luminous gas that shines faintly on the southeastern edge of the Large Magellanic cloud. It can be seen only from the Earth's southern hemisphere.*** **Below:** ***Edward C. Pickering (1846–1919) was Director of the Harvard Observatory from 1876 until his death, and masterminded the famous Draper Catalogue of Stellar Spectra, which is still in use today.***

Courtesy Celestron International

Others soon got into the act. The sun made its photographic debut in 1845, and five years later, the stars Vega and Capella were recorded. Astronomers were amazed at the startling possibilities.

They began to photograph eclipses, and regularly record sunspot activity on the face of the sun. They made photographic maps of the sky and used them to take accurate measurements of the stellar positions and motions.

In 1872, the first photograph was taken of the spectrum of Vega. It wasn't long before similar photos had accumulated by the hundreds. The variety of lines were almost as numerous as the photos themselves. Some appeared quite wide, while others were remarkably thin. Some were very dark, while others were barely visible. And some spectra hardly showed any lines at all, while others had thousands.

SEARCHING FOR RHYME OR REASON

Could any sense be made of this maze of lines? A remarkable and tireless young astronomer named Annie J. Cannon was determined to find out.

Cannon had become fascinated with spectra while a child in Dover, Delaware. One of her greatest delights was to watch the multicolored images cast by sunlight streaming through the glass prisms of a giant candelabrum hanging in her parents' house. Once her childhood days were over, Cannon entered Harvard College, and was surrounded by the world's finest collection of stellar spectra.

Under the direction of Professor Edward C. Pickering, director of the Harvard College Observatory, Cannon studied these stellar spec-

© Astronomical Society of the Pacific

tra in great detail. In a task analogous to grouping the books of an immense library, she painstakingly catalogued the spectra of countless stars across the heavens.

In time, she showed that the spectra of many stars were similar. She arranged them in a continuous sequence of broadening and fading lines. In fact, they were so similar that any one spectrum differed imperceptibly from the one immediately before or after it.

Cannon's findings—the results of her lifetime of research—the spectral classifications of a quarter of a million stars, were published in nine volumes by the Harvard College Observatory; they are still used nightly by astronomers around the world.

Just after the turn of the century, two other astronomers—Ejnar Hertzsprung and Henry Norris Russell—independently found an intriguing relationship. They discovered that the patterns of spectral lines indicated not only the chemical composition of the stars, but also their intrinsic luminosities as well.

On a very simple graph, the two men showed that hot blue stars were quite luminous, while cool red stars were often faint. They also concluded that one look at the spectrum of a star could place its position on the "Hertzsprung-Russell" diagram, and tell how brilliant and large the star was. By comparing how bright a star appeared in the sky with its newly found actual luminosity, astronomers could immediately calculate its distance.

It was, indeed, a remarkable century and a half. Technology had exploded, and telescopes and astronomical tools had become extremely powerful. Astronomers around the world began to peer more deeply into the Milky Way, and toward the faint, fuzzy nebulae that populated the universe.

Annie J. Cannon (1863–1941), an American astronomer who made an outstanding contribution to astronomy through her classification of the spectra of more than a quarter-million stars.

UPI/Bettmann Newsphotos

chapter six

ISLAND UNIVERSES

Until the early years of the twentieth century, astronomers had only the vaguest of clues about the true nature of the Milky Way. The best guesses came from the star hunters of the late nineteenth and early twentieth centuries who counted stars and estimated its size and shape. They concluded that the Milky Way was a lens-shaped system of stars 23,000 light years long and 6,000 light years thick, with the sun (and Earth) near its center. What could be more simple?

But one observation continued to make astronomers uneasy. They had noticed around the sky globes of thousands, even millions, of stars they called "globular star clusters." What was so strange was that these globular clusters appeared in only the half of the sky visible to the Earth's Northern Hemisphere during summertime.

Why these great star clusters should be distributed so

The horsehead nebula, so-named for its shape, is a dark nebula located in the constellation Orion. Consisting of a cloud of nonluminous interstellar matter, it is 300 light years from Earth.

Above: Henrietta Leavitt (1868–1921) spent most of her career at the Harvard College Observatory. Her work on variable stars of the Small Magellanic Cloud led to a determination of cosmic distances. Right: Harlow Shapley (1885–1972) began his college career as a journalism major until he began to ponder the universe. He became a pioneer of international collaboration in astronomy.

unevenly around the sky bothered astronomers around the turn of the century. One of whom was Harlow Shapley.

Shapley realized that the mystery of the globular clusters might be solved if he could measure their distances. His only problem was that no one had ever figured out how to do that.

THE COSMIC YARDSTICK

Shapley knew of a recent discovery made by Harvard College astronomer Henrietta Leavitt that might provide some help. For years, Miss Leavitt had been watching certain "variable" stars in a stellar grouping visible from Earth's southern hemisphere, known as the Small Magellanic Cloud. The stars, known as "Cepheid variables," appeared to change their brightnesses in a very regular way.

As she monitored their variability, Leavitt discovered an interesting trend. The Cepheids that appeared faintest to her varied in brightness most rapidly, while those with slower periods were the most luminous.

What this "period-luminosity" relationship meant, she had no idea. But astronomers soon recognized that, since the Cepheids of the Small Magellanic Cloud were all the same distance from us, their periods must also be related to their actual luminosities, a fact that could be used to calculate their distances, no matter where in the universe they appeared.

Having discovered this relationship, Shapley realized that in order to learn the distances of the globular clusters, all he had to do was to find Cepheid variables in them and

In this wide angle photograph of the Milky Way, thousands of stars and star clusters can be seen. Blocking our view of more distant stars are dark clouds of dust.

then measure their periods to determine their luminosities.

After hunting around the sky with the new sixty-inch (1.5-meter) telescope on Mt. Wilson in the San Gabriel Mountains near Pasadena, California, Shapley met with success. His results were staggering. One cluster, M13 in the constellation Hercules, was 36,000 light years away; most were even farther.

Several years and ninety-three globular clusters later, Shapley had extended his distance scale to nearly a quarter million light years. And then he made a remarkable conclusion: The Milky Way seemed to be a disk ten times larger than anyone had ever dreamed possible. In addition, the globulars formed a spherical halo above and below its plane, and were centered about a point tens of thousands of light years from us.

But something still didn't look right. Why should our tiny sun and solar system lie in the very center of the Milky Way, he wondered, while dozens of massive star clusters hover around some random point near its edge? It just doesn't make sense.

So in 1917, Shapley proposed a different approach. He believed it would be more reasonable if the massive globular clusters orbited the center of the Milky Way, while our tiny and insignificant sun lay out near its edge.

With this bold stroke of genius, Shapley did what Copernicus dared to do some four and a half centuries earlier: He removed our sun and Earth from the center of the universe. And, for his remarkable discoveries, Harlow Shapley became one of the most famous astronomers in the world.

***Right:** Shapley, seated here at his famous rotating desk, spent much of his career at the Mt. Wilson Observatory. From his work with globular star clusters, Shapley first determined the size of our Milky Way. **Far right:** Edwin Hubble (1889–1953) used telescopes at the Yerkes, Mt. Wilson, and Palomar observatories to catalogue and classify galaxies and measure the expansion of the universe.*

© Astronomical Society of the Pacific

UPI/Bettmann Newsphotos

THE ACCIDENTAL ASTRONOMERS

Harlow Shapley became an astronomer through an unusual twist of fate. Hoping for a career as a newspaper writer, Shapley left his home in Missouri to attend the renowned School of Journalism at Columbia University. When he arrived, however, he received a shock: The opening of the school had been put off for a year.

"So there I was," he recalled, "all dressed up for a university education and nowhere to go. 'I'll show them' must have been my feelings. I opened the catalogue of courses and got a further humiliation. The very first course offered was a-r-c-h-a-e-o-l-o-g-y, and I couldn't pronounce it! (Though I did know roughly what it was about.) I turned over a page and saw a-s-t-r-o-n-o-m-y; I could pronounce that—and here I am!"

Shapley wasn't the only astronomer of the day to enter the profession by way of the back door. Another was Edwin P. Hubble.

Hubble, like Shapley, was from Missouri. He was a brilliant, athletic man—a boxer who was so good that he was considered a legitimate contender for the heavyweight championship of the world.

While Hubble began his career as a lawyer in Kentucky, he soon realized his interests lay not in practicing law, but in solving the mysteries of the universe. "Even if I were second- or third-rate," he later recalled, "it was astronomy that mattered." So in 1919, he packed his bags and headed west to join the staff of the Mt. Wilson Observatory.

HIS WORK BEGINS

The many curious nebulae found scattered among the stars fascinated Hubble. By this time, astronomers had learned that some were clouds of hot glowing gases, and others were huge collections of stars.

Some suggested that these stellar groups were relatively nearby and were part of our Milky Way system. Others, like William Herschel nearly a century and a half earlier, argued that they were instead giant Milky Ways themselves—island universes—very far from Earth. No one knew exactly what they were, but Hubble was determined to find out.

Night after night, Hubble stood at the focus of the sixty-inch (1.5-meter) telescope, the smoke from his pipe swirling up and out of the great dome above. He photographed nearly every nebula that passed his gaze and soon realized that each nebula he saw fell into one of three distinct types. Most were shaped like a football, and he called them "elliptical" nebulae. Some appeared like a whirling pinwheel; he catalogued those as "spiral" nebulae. And some had no definite shape at all—the "irregular" nebulae.

Next, Hubble wanted to measure their distances. He realized that he could use the giant telescopes to find Cepheid variables within these nebulae. If so, he might be able to do what Shapely did with the globular clusters, and measure their distances once and for all.

For several years he observed variable stars in an irregular system known as NGC 2822. They looked like Cepheids, but he couldn't be sure. The sixty-inch (1.5-meter) telescope just

Courtesy Celestron International

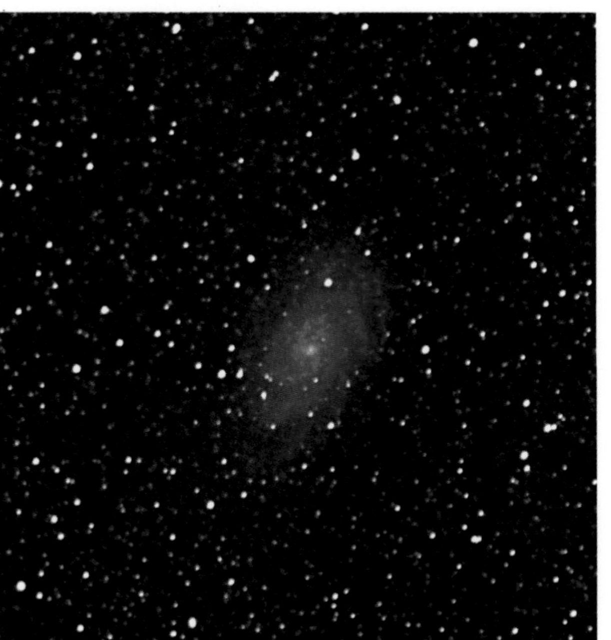
Courtesy Celestron International

wasn't big enough to tell. So, in 1924, Hubble aimed the new 100-inch (2.5-meter) instrument toward the outskirts of the Great Spiral in Andromeda.

Island Universes

This was a telescope beyond anyone's dreams. The 100-inch- (2.5-meter-) diameter mirror took six years to grind into a proper shape, and could gather nearly three times more light than the sixty inch (1.5 meter). Hubble knew it could reveal secrets the smaller scope could only hint at.

He was not disappointed. His first photos were so clear that he could pick out star clusters, gas clouds, and even some individual stars within the nebula. And, much to Hubble's delight, one of these stars was indeed a Cepheid.

It grew brighter and dimmer over a period of a month, and Hubble calculated that it was 7000 times more luminous than the sun. For the star to appear as faint as it did, Hubble reasoned, the spiral in which it was embedded must lie nearly a million light years away.

A million light years! There was no doubt any more. The Great Spiral in Andromeda was far beyond our own Milky Way. It was itself an island universe—a galaxy. Now Hubble was really intrigued, and he continued to photograph galaxy after galaxy to determine their distances.

Left: ***The Eagle Nebula, known also to astronomers as M16, is a huge cloud of hot gas illuminated from within by a cluster of newborn stars. Thick clouds of dark dust obscure this cluster from view and are seen here in silhouette. Spiral galaxies like this are found throughout the universe. It is only during this century that astronomers have recognized them as islands of hundreds of billions of stars far beyond our own Milky Way.*** **Below:** ***The Great Spiral Nebula in Andromeda was one of many similar nebulae scattered around the sky. During the nineteenth century, astronomers debated the nature of such spiral nebulae.***

Courtesy Celestron International

MEANWHILE IN ARIZONA . . .

Meanwhile, at the Lowell Observatory in Flagstaff, Arizona, another astronomer, Vesto M. Slipher, was hard at work on another problem. For fourteen years, he had studied the spectra of these nebulae, and found that none of their spectral lines appeared where they should. Those in the Great Andromeda Spiral all seemed to be shifted slightly toward the blue end of the spectrum, while nearly every other nebula showed its lines shifted toward the red end. Slipher knew exactly what this meant.

It was a principle first suggested in 1842 by the Austrian physicist Christian Doppler. Doppler noted that a train whistle sounds higher in pitch while the train approaches, but drops as the train recedes. Six years later, the French physicist H. Fizeau made the connection to light. He showed that light behaves the same as sound—not in pitch, but in color—and that an object's spectral lines would shift toward the blue as the object approaches, and toward the red as it recedes. This shift was known as the Doppler-Fizeau Effect or, more commonly, the Doppler Shift.

Slipher knew from its Doppler Shift that the Great Andromeda Spiral must be rushing toward us at a speed of 190 miles (304 kilometers) per second. Nearly all other galaxies he measured were receding from us, some as fast as 1,100 miles (1,760 kilometers) per second.

© Astronomical Society of the Pacific

Right: ***Vesto M. Slipher (1875–1969) was Director of the Lowell Observatory from 1916-1952. He was the first to recognize the redshifts within the spectra of distant galaxies, and provided a key element for understanding the evolution of the universe.*** **Below:** ***Percival Lowell (1855–1916) was a noted businessman, Orientalist, diplomat, and author. He built the Lowell Observatory in Flagstaff, Arizona, and became a full-time astronomer at age thirty-nine. He is seen here at the eyepiece of his giant telescope.***

American Museum of Natural History

In 1929, Hubble measured the distances to twenty-four of Slipher's galaxies, and discovered a strange pattern. The farther away a galaxy was, the faster it was moving away from us. If it was twice as far away, it was moving twice as fast. If ten times farther away, it was going ten times faster. This was too strange to be a coincidence. It had to mean something. But what?

OUR EXPANDING UNIVERSE

One afternoon, as he sat by a crackling fire in his home, Hubble puffed intently on his pipe while preparing for a fishing trip to the Rocky Mountains. Even his favorite pastime could not hold his attention, for his mind kept returning to the mysterious receding galaxies.

What was happening to these galaxies? Time after time he reviewed it in his head, but the only explanation he could imagine was that all the galaxies in the universe were rushing away from each other. The universe was expanding at an incredible pace.

After a relaxing trip to the Rockies, Hubble and his assistant, Milton Humason, put the 100-inch (2.5-meter) telescope to work on the problem. They were now looking at galaxies thirty-five times farther than those Slipher studied. These galaxies were so faint that a single photograph often required an exposure of ten successive nights to gather their light.

UPI/Bettmann Newsphotos

Edwin Hubble peers through the finder-scope of the forty-eight-inch (1.2-meter) Schmidt telescope on Palomar Mountain. Through his surveys of the heavens, Hubble found that the galaxies are rushing away from each other as the universe expands.

They studied galaxy after galaxy, and each showed the same phenomenon: The farther away it was, the faster it was receding. This "velocity-distance" relationship seemed to work everywhere they looked.

Hubble knew he had discovered a new cosmic "yardstick"—a method of measuring distances across the entire universe. All he had to do was photograph the spectrum of a galaxy, measure its "redshift," and he would automatically know exactly how far away the galaxy must be.

By 1935, Hubble and Humason had measured 150 new velocities. The fastest of all galaxies was moving at nearly 26,000 miles (41,600 kilometers) per second—one seventh the speed of light—and must, therefore, lie at the enormous distance of 240 million light years from us!

IN THE BEGINNING...

The most obvious questions on astronomers' minds were: Where are all these galaxies going? Where are they coming from? And, how long have they been flying apart? "If the universe is expanding," Hubble wrote, "it may finally be possible to determine the nature of the expansion and the time at which the expansion began—that is to say, the age of the universe."

And that's just what astronomers tried to do. They theorized that, at the beginning there were no galaxies, stars, or planets—no life or civilizations. All that existed was a tiny, but radiant, fireball. And then, they believed, two billion years ago, there was a colossal explosion.

Everything in the universe was hurled outward with a tremendous flash. Within the first few seconds, the universe cooled to a few billion degrees. Fundamental particles, such as protons, neutrons, and electrons, began to form. The forces we know today—the strong nuclear force, the weak electromagnetic force, and the gravitational force—separated from matter. As the universe expanded it slowly cooled. Clumps of matter coalesced into galaxies, still flying outward from the blast. Within each galaxy stars, and possibly planets, formed.

What existed before this Big Bang, no one could imagine; however, it seemed reasonable to many astronomers that this is how the universe began. Others weren't satisfied with the idea of such a beginning. They argued that something had to be there before, and they proposed another scenario to explain the receding galaxies.

They suggested that the universe had always existed and always will. As galaxies receded from one another, new matter was continually coming into being to fill the void. This Steady State theory seemed to defy the principle that matter could be neither destroyed nor created. Its supporters argued that only a tiny amount of new matter would be needed—a single hydrogen atom per year per cubic mile would do the trick. While many scientists were bothered by this, they had no proof that it was wrong.

Other probems lingered, as well. For example, why did every galaxy that astronomers saw seem to be smaller than our own Milky

FPG International

Engineers assemble the framework of the giant 200-inch (five-meter) Hale Reflector on Palomar Mountain. When this telescope was opened in 1948, it weighed nearly 150 tons (135 metric tons), yet it was so finely balanced that it could be moved with a one-sixteenth horsepower motor.

FPG International

***Left:** At the top of the giant Hale Reflector, astronomers rode inside this tube as the telescope tracked the stars. They would sit patiently for hours at a time, guiding the instrument while taking long-exposure photographs of the sky.* ***Below:** The Great Spiral in Andromeda was shown in this century to be not a cloud of gas and dust within our own Milky Way, but a separate galaxy of hundreds of billions of stars located two million light years away. Still, it was recognized as one of the nearest galaxies to us.*

Way? Why were the globular clusters in the Andromeda Galaxy considerably smaller than those of our home galaxy? And, if the Big Bang occurred only two billion years ago, why does the geological record of our planet suggest that the Earth was twice as old as the universe itself? Serious problems indeed!

Courtesy Celestron International

A BIGGER AND BETTER UNIVERSE

In 1948, a new telescope opened its eye on the universe: the 200-inch (five-meter) Hale reflector on Palomar Mountain, near San Diego, California. With this telescope, astronomers could peer billions of light years into space, and see galaxies rushing away at speeds of nearly 90,000 miles (144,000 kilometers) per second—nearly half the speed of light.

Astronomer Walter Baade began to learn that the Cepheid variable stars used for measuring distances came in two basic varieties, not one. And when he reviewed the distance scales of Hubble and Shapley, he discovered they needed to be revised.

His announcement in 1952 shocked the world: The Andromeda Galaxy was not a million light years away after all, but instead was two million light years away. And all other distances and sizes ever measured had to be increased as well. This meant that galaxies and globular clusters were actually larger and farther away than astronomers had thought, and that our Milky Way was just a medium-sized galaxy after all. Since the galaxies were now spread farther apart, the age of the universe must be increased from two billion years to perhaps five or six billion, considerably older than the age of the Earth.

The public may have been surprised by the news, but there was one man who had seen it coming. He had never viewed the universe through a giant telescope, nor had he studied the spectra of photographs of distant galaxies. His only view came from the power of his mind. His name was Albert Einstein.

$R_{ik} = 0$

chapter seven

THE UNIVERSE OF DR. EINSTEIN

Albert Einstein (1879–1955) looked at the universe in a free and uninhibited way. He asked questions that many might have considered trivial, and sought to answer them through the language of the universe—mathematics.

It was the 1890s. Through the hills of Northern Italy strolled a rather thin, long-haired, high-school dropout. As he walked the road toward Pavia, he gazed out across the land and became intrigued by the beams of sunlight glistening off church windows, lakes, streams, and mountain tops. He began to wonder: How would the world appear if I could ride a beam of light? Why does light travel at the speed it does? Why not faster or slower? If I rode a beam of light and encountered another coming toward me, would it appear to go twice as fast as I am?

As he walked, his mind raced with questions that challenged the obvious, questions that might never even occur to the average person. But then, young Albert Einstein was not your average young man.

THE TROUBLEMAKER

Several years earlier, at the tender age of twelve, Einstein had become fascinated by science after reading a book entitled *People's Book of Natural Science.* From that moment on, his intense curiosity about the world and universe around him often got him into trouble.

In school, his never-ending questions disrupted the highly regimented German classes, and his teachers became increasingly frustrated, finally telling him that he would be better off out of school.

So Albert took their advice and struck out on his own. And what a brilliant decision it was, for it was in the serenity of Northern Italy that Einstein experienced the happiest times of his young life. Here, he could think and ponder the universe freely without being criticized by his overly disciplined teachers.

THE NEW SCIENTIST

It was indeed a wonderful time, but it could not last forever; Albert soon found himself in need of a job. He returned to finish school in Switzerland, completed his studies in physics and mathematics, and eventually found work in a Swiss Patent Office, evaluating patent applications.

He thought it was a perfect job. Each day he finished his assignments early and had plenty of time to ponder the nature of the universe.

In 1905, Einstein received his doctorate from the University of Zurich and continued his work at the patent office. It was here that Dr. Einstein completed his theory to address the paradoxes that occur when traveling at very high speeds. He called it the Special Theory of Relativity.

THOUGHT EXPERIMENTS

In developing the theory Einstein enjoyed performing what he called a "gedankenexperiment," a thought experiment.

He imagined that a train was approaching him at 300 miles (480 kilometers) per hour and fired a missle toward him at 500 miles (800 kilometers) per hour. On the ground he would perceive the missile's speed at 500 miles per hour (800 kilometers) plus 300 miles (480 kilometers) per hour or 800 miles (1280 kilometers) per hour. A passenger on the train, however, would clock the missile at only 500 miles per hour (800 kilometers), since that's how fast it moves away from him. It seemed reasonable enough.

But where Einstein had problems was not with trains and missiles, but with beams of light. He wondered what would happen if, instead of firing a missile, the approaching train had turned on its headlight. Would the beam come toward him 300 miles (480 kilometers) per hour faster than if the train were stationary?

That just can't be, he thought. If it were, then there must be some special place in the universe where the speed of light would correspond to that predicted by theory, while all others are somehow less special.

Einstein just couldn't accept that. After all, people had been trying to establish a cosmic "place of distinction" ever since the days of the Earth-centered universe. Einstein wasn't about to fall into that trap.

He concluded, instead, that if the speed of light were constant, as theory predicted, then everyone must measure it to be the same—regardless of their motion. And, since "speed" is just the distance something travels over a certain period of time, Einstein concluded that it must be the passage of time that is perceived differently.

As he thought about it, he discovered that other unusual effects would also occur. Suppose someone is riding a spaceship at nearly the speed of light, he imagined. Everything onboard appears perfectly normal to him.

But, he continued, I am standing on the Earth, and something quite different. To me, the Doppler Effect makes the spaceship appear bluer as it approaches and redder as it recedes. Not only that, but the faster it goes, the shorter and more massive it appears, and the more massive it appears, and the more slowly its onboard clock seems to run. Then, just as the ship attains the speed of light, I see its clock seem to stop running, the ship's length is reduced to zero, and its mass has grown to infinity!

While Einstein knew that two different perceptions of the same phenomenon defy common sense, he also realized that the universe was not designed to satisfy human senses. He concluded that neither view was wrong, just relative to the observer's position.

His Special Theory of Relativity addressed the paradoxes that occur at very high speeds.

The Bettmann Archive

While still a young man, Einstein made an indelible mark in modern science. By the age of forty-two he had published his special and general theories, and had been awarded the Nobel Prize for his work on the photo-electric effect.

It also showed that matter was just an extremely concentrated form of energy and that, under the right conditions, tremendous amounts of energy could be released from tiny bits of matter.

What About Gravity?

As remarkable as it was, the Special Theory left a lot unsaid. For example, it didn't account for the effects caused by gravity. Einstein explained those in his General Theory of Relativity.

He looked at it like this. If a bullet is fired horizontally in a gravitational field or in a spacecraft that is accelerating upward, its path will appear to bend downward. Isaac Newton thought of it as the attraction of gravity. The same must be true for a beam of light, Einstein thought. Light should be bent by gravity.

Einstein knew that a light beam always followed the shortest path between two points, but the only way this could happen was if he described gravity differently than Newton. One way was to consider space to be like a thin rubber sheet, on which a heavy ball was sitting. The massive ball distorts the two-dimensional sheet into a third dimension by creating a depression. Another ball rolling by might fall into the depression, or it might just change the direction that it is rolling.

An observer on the outside would perceive the ball's path to bend as the ball is "attracted" inward. In this way, Einstein saw gravity not as a force between two objects, but as a distortion of the very "fabric" of space. He linked the three dimensions of space and one dimension of time into a four-dimensional space-time.

Mathematically it was complicated, but Einstein explained it simply. "If you will not take the answer too seriously," he said, "and consider it only as a kind of joke, then I can explain it as follows. It was formerly believed that if all material things disappeared out of the universe, time and space would be left. According to the relativity theory, however, time and space would disappear together with the things."

Scientists soon devised experiments to see if Einstein's ideas were correct. They checked star positions during a total solar eclipse to see if starlight bent as it passed near the massive body of the sun. It did. They flew precise atomic clocks on jet aircraft to see if time slowed down at high speeds. It did.

In fact, all predictions made by Einstein's theories turned out to be correct. All, that is, except for one.

THE EXPANDING UNIVERSE

Einstein felt so strongly that the universe must be a static place, that when his equations suggested it was not, he inserted a "fudge factor" to make it static. When astronomers discovered that the universe was expanding in the 1920s, Einstein dug out his old calculations, and found his error. Disgusted with himself for having done such a thing, he called it "the biggest mistake I ever made."

When the equations were corrected, they made more sense. They showed that it was not the galaxies that were flying apart from one another as it might seem. Instead it was the very fabric of space and time that was actually stretching, and was carrying with it the galaxies.

It also seemed that there might be two possible types of universe. One was a universe that went on forever in all directions; an open universe, which would never stop ex-

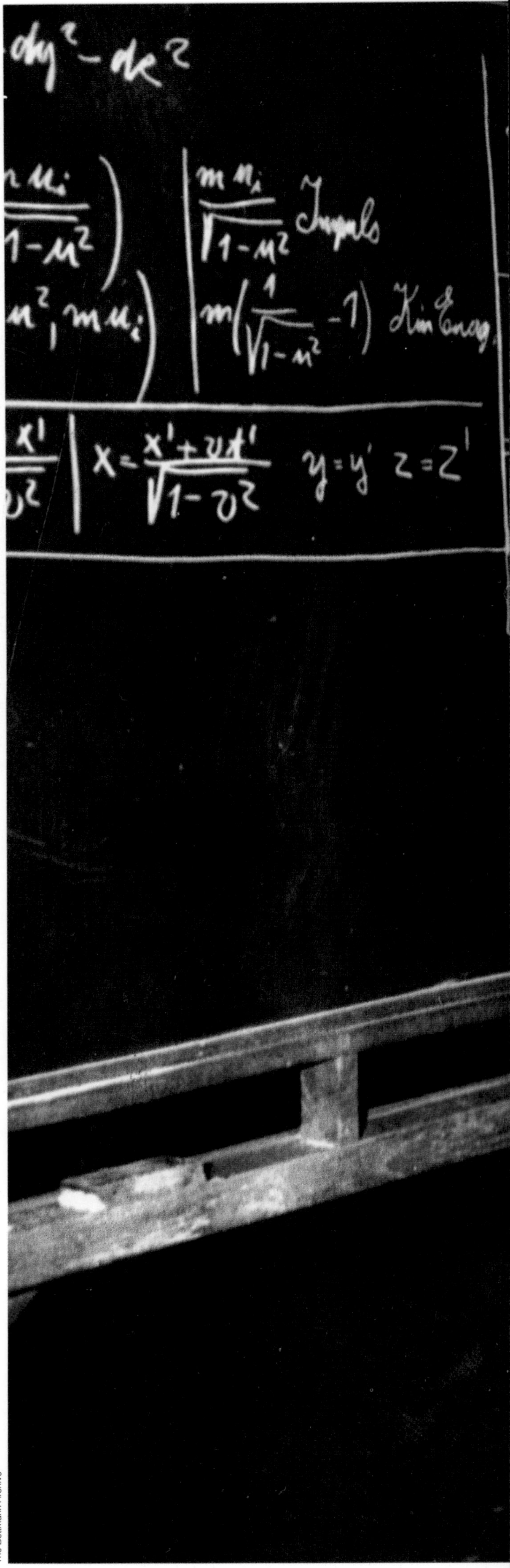

The Bettmann Archive

An extremely popular and engaging speaker, Einstein is seen here in December 1934 as he delivered a talk for the American Association for the Advancement of Science at the Carnegie Institute of Technology.

Right: ***Dr. Einstein and his wife sailing for home on the*** **S.I. Celtic. Opposite page:** ***Though Einstein's tools were a pencil and paper, he always enjoyed peering through telescopes at the wonders of the universe.***

The Bettmann Archive

panding. A journey in a straight line through this universe would never end.

Another possibility was a closed universe. This was one in which a straight journey would take us out past distant galaxies and ultimately bring us back to our home galaxy—from the opposite direction.

Einstein was purely a theoretician, but he always enjoyed seeing the equipment that could make such fantastic discoveries. Once in 1930, during a visit with Edwin Hubble at the Mt. Wilson Observatory in California, Doctor and Mrs. Einstein were given a tour of the facility. As they passed beneath the giant 100-inch (2.5-meter) telescope, it was explained to them that this instrument was used to determine the structure of the universe. "Well, well," nodded Mrs. Einstein, with a smile. "My husband does that on an old envelope!"

WELCOME TO AMERICA

During the 1940s, Hitler came to power and the terror of Nazi Germany spread throughout the entire European continent. Since his youngest days, freedom was very important to Albert Einstein. And so he and his wife soon fled to Princeton, New Jersey, where he could carry out his research without fear.

Americans were soon impressed with the brilliance and warmth of the aging scientist and sought out his opinions on everything from science to religion to politics.

Once he was asked the secret of his long and successful marriage. "When Elsa and I were married," explained Einstein, "we agreed that she would make all the minor decisions and I would make all the major ones." Then he paused for a moment and continued with a sparkle in his eye, "It's just that there have never been any major decisions to be made!"

A NEW UNIVERSE

The discoveries he made were more than enough. Through the marvelous language of mathematics, Einstein described a universe far more elegant than anyone had ever imagined. Not bad for a young man who would "never amount to anything."

In 1955, Einstein died, and his brilliant mind, unending curiosity about the mysteries of nature, and sparkling sense of humor were silenced forever. Fortunately, the precious gift he left to humanity will never be lost; an all-too-brief look inside the remarkable, and often paradoxical, universe of Dr. Einstein.

UPI/Bettmann Newsphotos

chapter eight

NEW EYES

Anyone passing this field in northern New Jersey during the summer of 1931 would surely have looked twice at the odd-looking contraption sitting there. It was 100 feet (thirty meters) long and made of thin brass pipes and wooden two-by-fours clamped together. It looked for all the world like the frame of a long, thin house, except that it sat on four automobile tires, which fit neatly into a circular track. When the contraption turned, it creaked and rattled as if it would fall apart any second.

With giant radio telescopes like this one at Stanford University, scientists can "see" radio radiation that is invisible to the eye. This radiation is generated by such things as interstellar clouds of gas and dust, exploding stars, and black holes.

What passersby didn't see, however, were the wires that led from the contraption into a nearby laboratory, where a thin, balding man named Karl Guthe Jansky sat in front of an amplifier with earphones over his ears.

Jansky was an engineer for the Bell Telephone Laboratories in Holmdel, New Jersey. He was using this peculiar radio antenna to track down sources of static that could possibly

Courtesy Bell Laboratories

Courtesy Bell Laboratories

In the 1930s, Bell Telephone engineer Karl Jansky (1905–1950),* left, *built this unusual rotating antenna in a field near Holmdel, New Jersey* (far left), *and discovered radio waves coming from the sky. His work led to the birth of an entire field of study: radio astronomy.

disrupt shortwave radio and telephone communications.

STRANGE NOISES

During his daily experiments, Jansky managed to identify a number of possible sources of noise, but there was one he couldn't explain. It seemed to be coming from the sky.

Jansky had heard the radio noise of thunderstorms many times, but this was different. This was a weak, steady hiss that occurred every day, moved slowly across the sky, and disappeared just after sunset. The noise also seemed to "rise" four minutes earlier each day. Within a week, its starting time had advanced by nearly half an hour.

Soon, Jansky started getting up in the middle of the night to drive to his lab so he could listen to the mysterious static. Whatever it was, he hadn't a clue. And then, he hit upon an idea. After checking some astronomy books in his local library, Jansky discovered that the stars themselves rise in the east and set in the west four minutes earlier each day—all because of the motions of the Earth.

Jansky knew the answer almost immediately. The mystery signals were a mystery no longer. They were coming from the stars.

Over the next few months Jansky tried to isolate the direction of the noise. In the spring of 1933, he published his findings. "The radio waves may be coming from the center of the

***Right:** Karl Jansky was instrumental in detecting radio waves coming from the center of the Milky Way; however, most astronomers thought these waves were nothing but noise.* ***Far right:** Otto Struve was the astronomer to whom Grote Reber took his sky charts and maps for interpretation.*

© Astronomical Society of the Pacific

UPI/Bettmann Newsphotos

Milky Way," he wrote, "or from the constellation of Hercules."

Jansky's discovery of radio noise emanating from the depths of space caused quite a sensation. One New York radio station even broadcast the signals coast-to-coast. Families around the nation huddled together by their radios and listened in utter amazement as the announcer proclaimed: "Our broadcast tonight will break all records for long distance. We shall let you hear radio waves from somewhere among the stars."

While the public's imagination was stimulated, that of astronomers was not. Most felt that the noise was just that, noise, and studying it served no useful purpose. So in April 1937, Jansky abandoned his plans to continue his work with an even larger antenna.

A New Antenna

Not everyone was so pessimistic. In Wheaton, Illinois, a ham radio operator named Grote Reber became intrigued at the prospects of building a huge dish to capture and study these cosmic radio waves. During the hot summer nights he worked on his antenna; it soon towered above the trees. Neighbors peeked curiously over the fence at the strange goings-on in the Reber back yard.

In August 1937, Reber finished the antenna and began his work. Night after night, week after week, he listened from his basement control room, but heard nothing. He checked his wiring, changed his amplifiers, and tuned to longer and longer waves—three inches (7.5 centimeters), four inches (ten centimeters) . . . , two feet (sixty-one centimeters), three feet (ninety-one centimeters). Still he heard nothing.

Then one night in October, in utter desperation, Reber tuned to the longest radio waves he could imagine. Suddenly, the needles in his meters came alive. "A signal!" he thought. "No wonder I haven't heard anything. These waves are six feet (two meters) long!" Reber sat throughout the night, mesmerized by the hiss he had just discovered.

The First Radio Astronomer

By day, Reber went to work designing home radio receivers in Chicago. By night, he worked as the world's first "radio astronomer." He would often arrive home after a full-day's work, eat dinner, and sleep until midnight. Then he would rise and head straight to his basement lab where he would spend the night taking readings from the sky. Around 6

A.M. each morning, he would shut down his equipment, drink some strong, hot coffee, and head off for another long day in the Windy City.

Once Reber began to interpret his sky charts and maps, he took his work to the Yerkes Observatory in Williams Bay, Wisconsin. In the office of Otto Struve, the director of the observatory, Reber excitedly spread out his maps. They looked like contour maps of the Earth, with each line indicating a different signal intensity.

Reber believed that his results seemed to confirm Jansky's discovery. Dr. Struve stared with curiosity at the charts. Soon other astronomers began to gather. No one had a clue what could be producing these radio waves, but everyone was fascinated.

In 1940, Reber published his first results, and immediately set out to refine his equipment and improve his radio map of the heavens. By scanning his dish north and south and allowing the Earth to carry it east and west, Reber was able to map the entire sky.

When he did, he found that most of the signals were coming from the plane of the Milky Way. Some emitted particularly strong broadcasts. "One is toward the center of our galaxy, in the direction of Sagittarius," he wrote. "Others seem to be placed in arms

UPI/Bettmann Newsphotos

Otto Struve, a Russian-American astronomer, was director of the Yerkes Observatory in Williams Bay, Wisconsin, when Grote Reber first presented his celestial radio maps.

projecting from our galaxy—perhaps in spiral arms of the type seen in other galaxies."

Reber had discovered a fact that would be used by astronomers to map the structure of the Milky Way galaxy: Radio emissions arose naturally from the gases of interstellar space. More importantly, however, he showed a skeptical world that the sky was emitting important information which, until then, had been totally ignored.

ONLY THE BEGINNING

Today we know that the universe glows in many different colors. What we see as light—the red, orange, yellow, green, blue, and violet that make up the visible spectrum—is only a tiny fraction of the radiation being emitted. Just as a piano keyboard extends over several octaves of sound, the entire electromagnetic spectrum extends over thirty "octaves" of radiation, with visible light making up less than one. Astronomers soon realized that trying to understand the universe from visible light alone was like trying to identify a fine piece of music played on a piano with only six functioning keys.

Knowing the radiation was there was one thing; capturing and studying it was something else. One problem was that most radiation never made it past our atmosphere to the ground. Light and radio waves got through just fine, but some, like ultraviolet, x-rays, and gamma rays, were blocked by our atmosphere. It's a good thing, too, or life on Earth could never exist.

Radio telescopes, such as these in California (below) and Texas (opposite page), have forever changed the way we study the heavens. The largest, in Arecibo, Puerto Rico, is so powerful that it can detect signals transmitted from the other side of our galaxy.

© Mike J. Howell/Envision

Another problem was that each type of radiation required a different kind of detector to "see" it. This equipment had to be taken where it could capture the radiation—to dry mountaintops, high into the atmosphere, or into space. Until the late 1930s, the technology to do this just didn't exist. But with the end of World War II, things began to change. New electronics and powerful detectors were developed. No longer were we bound to the surface of our planet. Balloons, high-flying aircraft, rockets, and satellites could now take our equipment to places visited in our dreams.

Almost overnight, the impossible became commonplace, and we began to see the universe through totally new eyes. What we've learned is spectacular.

PUZZLING NEW RADIO SOURCES

Today, radio telescopes are many feet wide. The largest is in Arecibo, Puerto Rico, and spans the length of 3⅓ football fields. Carved into the ground, this telescope is so powerful that it could detect signals transmitted by a twin dish on the other side of our galaxy.

In the five decades since the field began, radio astronomers have made remarkable discoveries. They have identified more than a hundred chemical elements and molecules among the gasses in space, traced the spiral pattern of our Milky Way galaxy, and watched the violent collisions of distant galaxies. They have teamed with optical astronomers to locate individual sources of radio waves throughout the universe. And many of their discoveries came about purely by accident.

One of the most exciting periods in radio astronomy was the decade of the 1960s. It began with the discovery of a peculiar object listed as the forty-eighth entry in the Third Catalogue of Cambridge Radio Sources. Its name was 3C48.

Optical astronomers first aimed their telescopes in its direction in 1960, but couldn't pinpoint the source of the radio emission. Then, in 1963, radio astronomers in Australia took advantage of a unique "occultation." They listened as the signals of a similar object—this one called 3C273—were blocked by the moon passing in front of it. From their precise timings, the scientists could now pinpoint the object's location, and give optical astronomers more accurate coordinates.

FPG International

Above: ***It was with the giant 200-inch (five-meter) Hale Reflector on top of Palomar Mountain that astronomers first found the bizarre quasi-stellar objects they named "quasars."*** **Right:** ***Quasar 3C 275.1, shown here in a false color photo, is the first quasar found at the center of a galactic cluster.***

It didn't take long to find the object with optical telescopes. It looked like a normal star, but its pattern of spectral lines was unlike anything ever seen. Was it made of chemicals we couldn't identify? Or, was there some bizarre new form of physics at work inside?

None of it made any sense, until Hale Observatories astronomer Maarten Schmidt had an idea. He decided to slide several lines of hydrogen toward the red end of the spectrum to see if he could get them to agree with the pattern showed by 3C273. Farther and farther he slid them until, suddenly, three lines clicked into place.

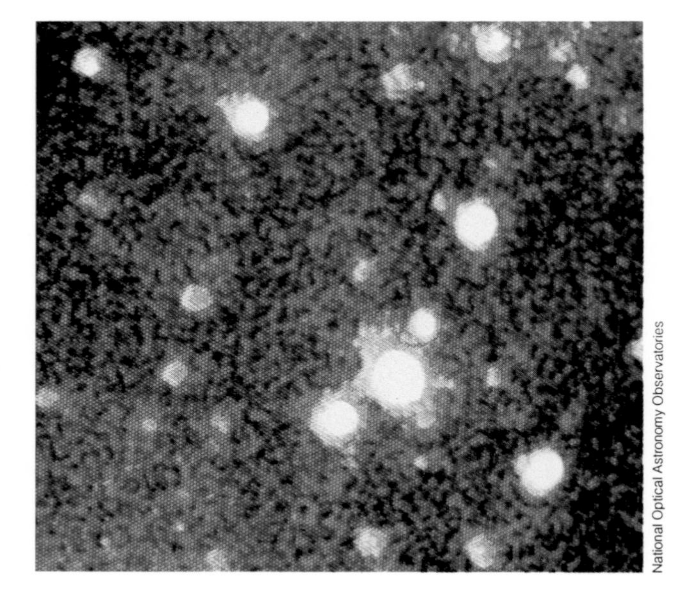

National Optical Astronomy Observatories

Schmidt now realized why the object's spectral lines had astronomers so baffled. They were all redshifted by nearly 16 percent. If I'm right, he thought, this object is rushing away from us at 69,000 miles (110,400 kilometers) per second, or 37 percent the speed of light and, according to the velocity-distance relationship Hubble had discovered, 3C723 must be located more than two billion light years away.

As if this weren't odd enough, the object's brightness was found to flicker wildly every few days, indicating that it must be extremely small, perhaps only a few times larger than our solar system.

The object was called a quasi-stellar object or "quasar" for short, and it created a nightmare for astronomers. Here was a relatively tiny object located at the visible edge of our universe. To appear as luminous as it did, it must be pumping out tremendous amounts of radiation, as much as a hundred entire galaxies. No one could imagine what energy sources were at work within.

Or, astronomers thought, maybe it wasn't tiny and powerful at all. Perhaps the quasar only appeared so luminous because it was nearby. But then, why did its spectrum show such an immense redshift? Did this mean that the velocity-distance relationship of Hubble was wrong?

In the ensuing years, more and more quasars were found. Most appeared farther away and more luminous than 3C273. And, as astronomers furiously continued their debate about the nature of the quasars, another discovery was being made across the continent.

FINDING THE BIG BANG

For many years, astronomers had debated whether the universe really began as a colossal explosion billions of years ago, or whether it was always here as the Steady State theorists claimed.

Since the 1940s, astronomers knew there was one way of settling the argument. If the universe began from an explosion of a superhot fireball, the fireball's radiation should still be there, cooled now to only a few degrees above absolute zero. Find the radiation, and the problem is solved. Unfortunately, the technology to do so was many years away.

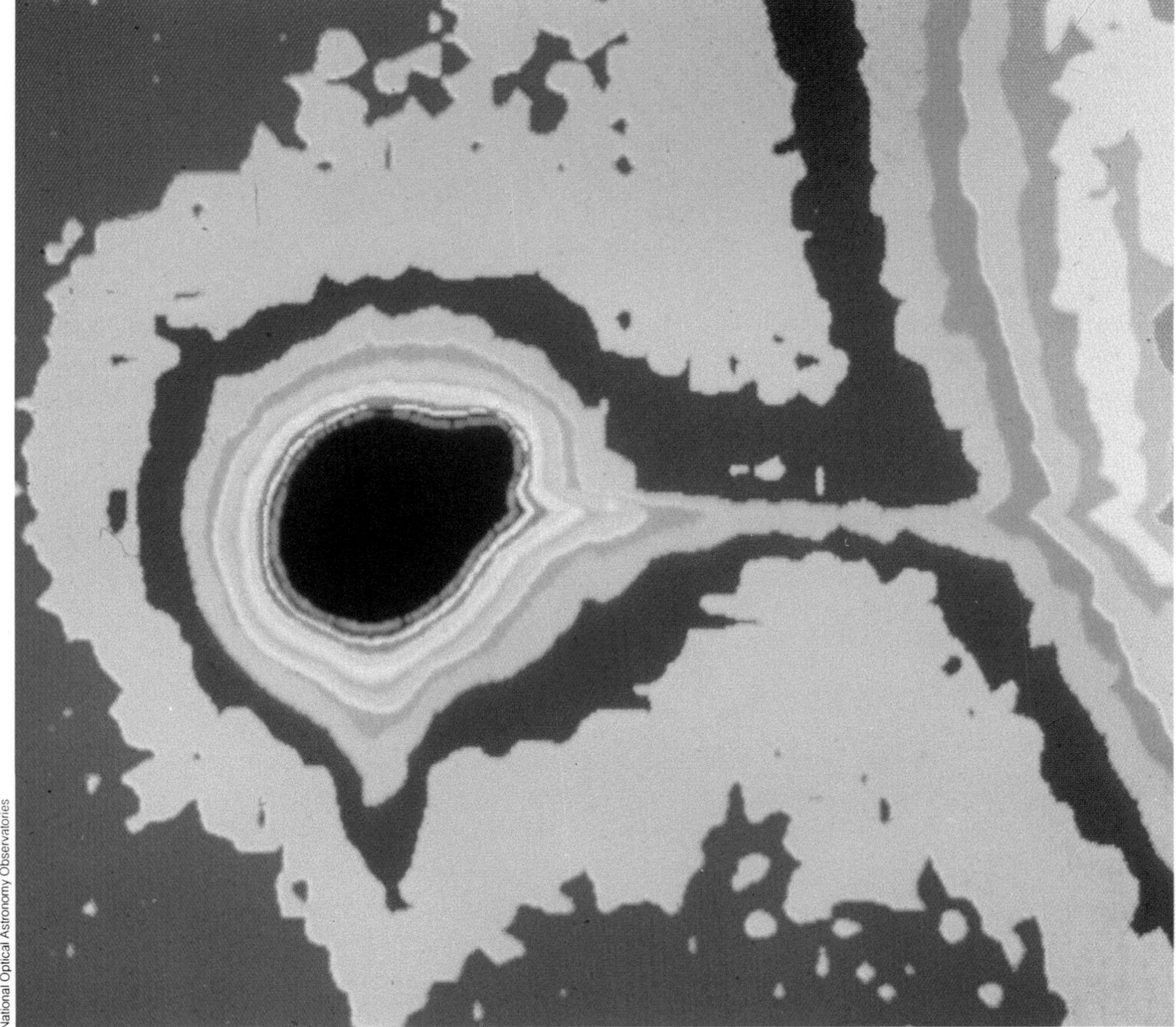

National Optical Astronomy Observatories

Some quasars seem to be connected to galaxies through luminous "bridges," as shown in this computer-enhanced image of the galaxy NGC 4319 and quasar Markarian 205. Conventional measurements place this quasar at a distance of 912 million light years away.

Bell Laboratories scientists Arno A. Penzias* (left) *and Robert W. Wilson* (right) *stand in front of the radio antenna they used to detect the radio emissions left over from the birth of the universe. In 1978 they were awarded the Nobel Prize in physics for their discovery.

It wasn't until the 1960s, with the debate still raging, that a team of physicists from Princeton University began building a detector to settle the debate once and for all. But, only a few miles away, again in Holmdel, New Jersey, two Bell Telephone physicists were making their own discovery.

Arno Penzias and Robert Wilson were trying to improve communications with artificial satellites when they found a faint, persistent static coming from the sky. It was different than the rising and setting radiation Jansky had found three decades earlier; this came from everywhere.

They tried to eliminate it, but failed. They had no idea what was wrong. When they learned of the work being done at Princeton, they knew immediately what they had found. It was the remnant radiation from the hot gasses of the Big Bang itself. In one historic and purely serendipitous discovery, Penzias and Wilson had found the remnant radiation from the birth of our universe.

More Mysteries

Two years later, another accidental discovery was made, this time by Cambridge University astronomer Antony Hewish and his graduate student Jocelyn Bell.

In July 1967, Hewish put Bell to work studying how radio sources "twinkled" as their beams were interrupted by particles blasted from the sun. Night after night, Bell sat in her lab and watched as data came in from around the sky. Then, around midnight on August 6, something peculiar caught her eye. The pen on the strip-chart recorder fluctuated wildly as something unusual passed over her antenna. Maybe it was caused by local radio interference, or maybe the telescope had caught a distant star in the act of flaring in brightness. Before she could determine the source of the interference, it faded from view.

Then, on November 28, the phone rang in Dr. Hewish's home. "It's back!" said Bell, on the other end. Only this time, it had sharp pulses that were regularly spaced at intervals of just over one second!

"It was hard to believe," recalls Hewish, "and I was not convinced until I actually saw the pen tracing similar pulses on the following day."

Immediately, the astronomers began to study the mysterious signal. They soon found that it was coming from a source smaller than Earth located a few hundred light years away. In addition, it was pulsing with an accuracy of a millionth of a second. Hewish wondered if it was possible they had detected a signal from an extraterrestrial civilization?

As they continued to study the signal, it became clear that the source was not an artificial signal being sent from an alien planet, but a natural signal emanating from an object never before seen by astronomers.

When their report was published in Febru-

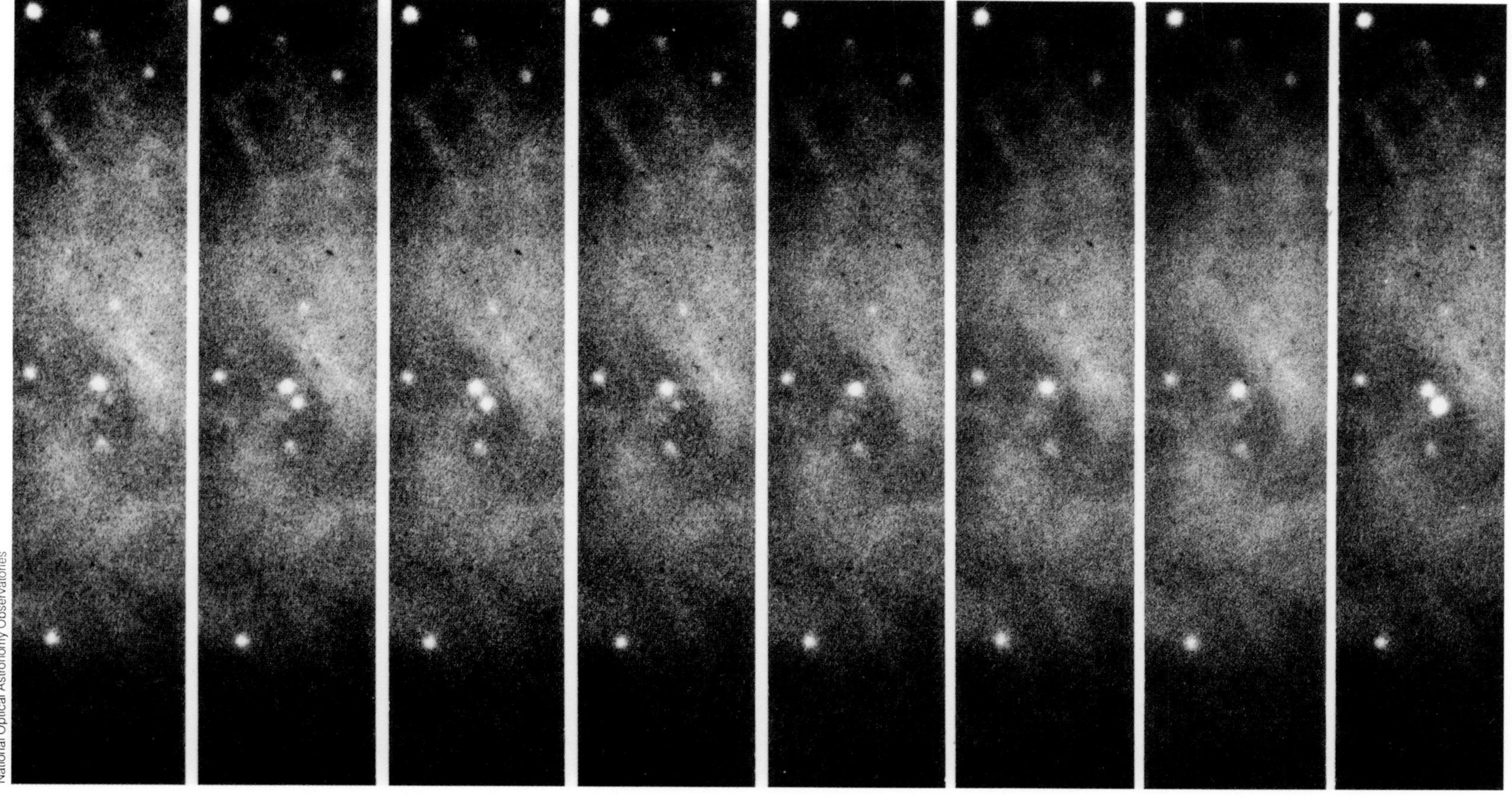

Inside the Crab Nebula, a remnant of an ancient stellar explosion, astronomers found a pulsar flashing in visible light. This sequence of photographs shows the pulsar NP 0532 as it pulsates with a period of 1/30 second.

ary 1968, it astonished the astronomical community. The object they discovered was a tiny, dense neutron star—the remnant of a violent stellar death—that spins wildly and funnels radiation out its magnetic poles like a lighthouse beacon.

Since Bell's discovery in 1967, hundreds of such "pulsars" have been found around the sky. Many flicker several times each second and one, discovered in 1982, pulsates at a rate of 600 times every second!

X-RAY EYES

During the 1960s, astronomers unlocked other "windows" of radiation. One was the high energy portion of the spectrum known as x-rays and gamma rays.

The study of these rays actually began in 1949 when scientists at the Naval Research Laboratory (NRL) found that the sun was emitting x-rays. All were blocked by our atmosphere; so in order to see them, astronomers had to launch their equipment high above the Earth with V2 and other sounding rockets.

Most astronomers felt that if x-rays were coming from other sun-like stars as well, they would be much too faint to see; which is why a discovery made by Riccardo Giacconi and his colleagues in 1962 came as such a shock.

With instruments carried into space by a sounding rocket, Giacconi found an unusually strong x-ray source lying in the direction of the constellation of Scorpius. Soon, others began to appear around the sky. No one knew what they were, or how their radiation was being generated.

Then, five years later, another surprise came. In the midst of the cold war between the United States and the Soviet Union, American satellite-borne detectors monitored the ground for violations in the nuclear explosion test ban treaty. What they found instead were bursts of energetic gamma rays coming from the sky.

Throughout the fifties and sixties, astronomers had peeked for a few hours into the marvels of the high-energy sky, and had discovered only a few dozen x-ray and gamma ray sources. But on December 12, 1970, things began to change.

On that day, a satellite named *Uhuru* was launched into Earth-orbit, carrying with it the most sophisticated x-ray detector ever made. With the satellite, Giacconi and his colleagues at American Science and Engineering, Inc. carried out the first complete survey of the sky in this mysterious radiation. They found some 400 separate x-ray sources around the sky. Still, questions remained: What was causing the x-rays? Where were they coming from? And, how many more were out there?

Then, in the late 1970s, a series of High-Energy Astrophysical Observatories (HEAOs) was launched into orbit to study the high-energy phenomena that had astronomers puzzled. The first of the series, *HEAO-1*, per-

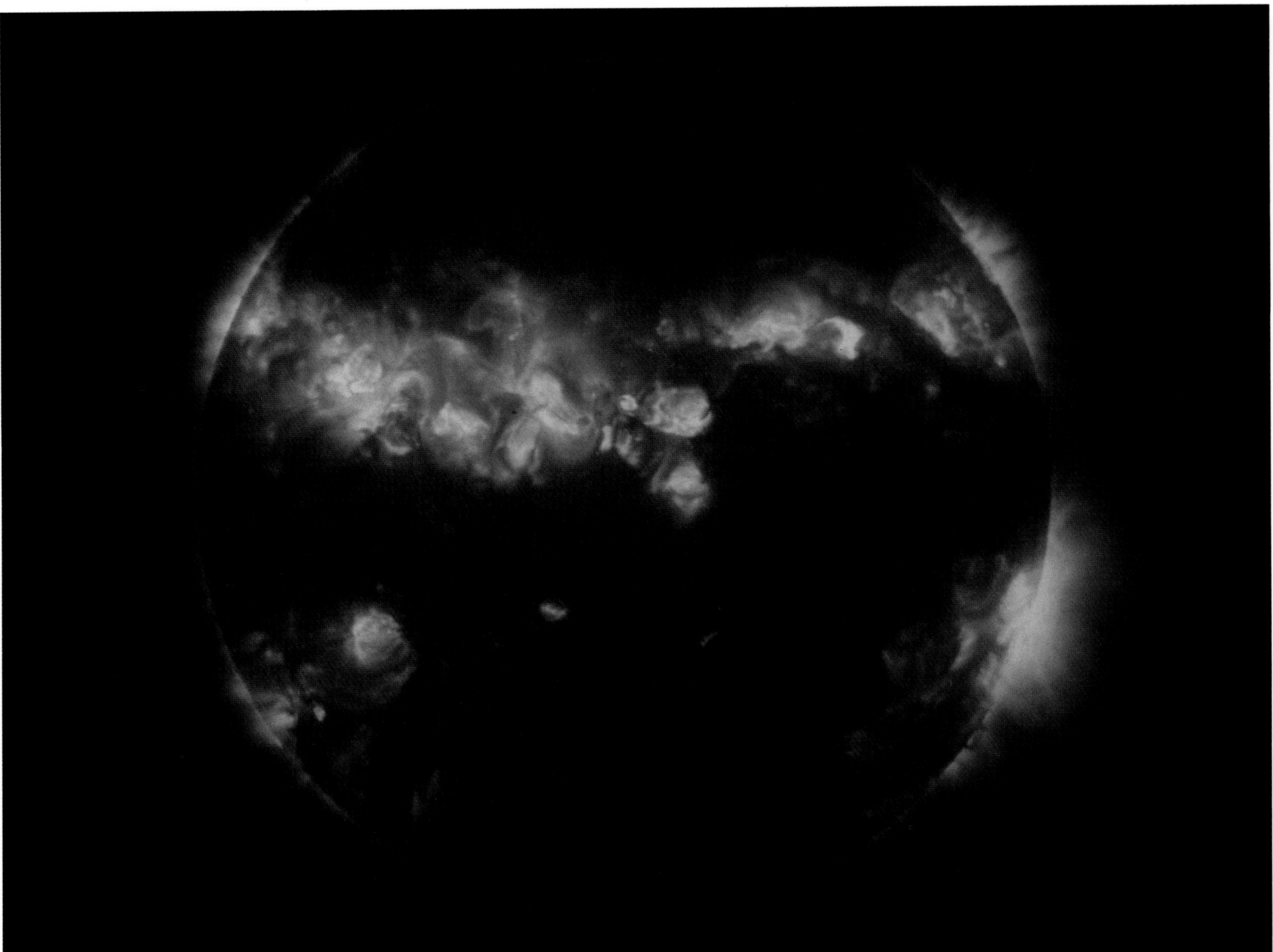

Courtesy SAO and IBM Research

Above: ***This x-ray image of a solar flare was taken by a Smithsonian Astrophysical Observatory/IBM telescope aboard a NASA sounding rocket.***

formed a more sensitive survey of the heavens, and searched for sources that might be varying in x-ray brightness. It was followed into orbit by *HEAO-2*, also called the Einstein Observatory. During its 2½ year mission, the Einstein Observatory catalogued thousands of new sources, studied those found by its predecessors, and learned that many stars of our galaxy are x-ray emitters.

It was soon followed into space by the third of the series, *HEAO-3*, which studied more energetic gamma radiation and high-energy particles from space known as Cosmic Rays.

UV Windows

Another window that opened during the 1960s was the ultraviolet window. Like other "new" astronomies of this time, ultraviolet studies required trips high above our atmosphere in balloons, rockets, and satellites. But unlike the others, few new sources of UV were ever found. Most were just hot, normal stars. So astronomers concentrated on studying the ultraviolet spectra of known stars and galaxies to learn their temperatures, densities, chemical make-up, and speeds.

In 1973, scientists launched the thirty-six-inch (ninety-one-centimeter) telescope on board the third Orbiting Astronomical Observatory (OAO). During its seven year mission, OAO studied the ultraviolet spectrum of the sun, other stars, interstellar material, and distant galaxies.

Five years later, the International Ultraviolet Explorer (IUE) of the National Aeronautics and Space Administration (NASA) and the European Space Agency (ESA) and the British Science and Engineering Research Council (SERC) carried an eighteen-inch (forty-six-centimeter) telescope to gather even more efficient data than OAO.

From their newly found data, astronomers learned that high-energy radiations were coming from some of the most violent and exotic phenomena in the universe. Some may be remnants of shattered stars or explosive jets of gas bursting from distant galaxies. Others may be emanating from hot stellar winds or

Courtesy Celestron International

***Left:** This photo of the Lagoon nebula indicates super-heated clouds of gas.*
***Below:** This near-infrared, false-color image is one of the first made of the rings of Uranus with a ground-based telescope.*

super-heated clouds of gas in space. Still others indicate the presence of tiny, supermassive objects pulling into them nearby gasses, and heating them until they glow in high-energy radiation. These may be the strange and forboding black holes.

Cosmic Heat

Just as ultraviolet, x- and gamma radiation tells about the hottest, most violent processes in the universe, infrared tells of the coolest objects known. With infrared telescopes, astronomers can see through shrouds of dust that hide newborn stars, probe the center of our galaxy, and study the cooler objects closer to home, such as the planets, moons, asteroids, and comets of our solar system.

Most infrared radiation is absorbed by water vapor within our atmosphere, so astronomers have made great efforts to place their instruments above the vapor. They have launched remote-controlled telescopes aboard balloons and rockets to gain brief glimpses of

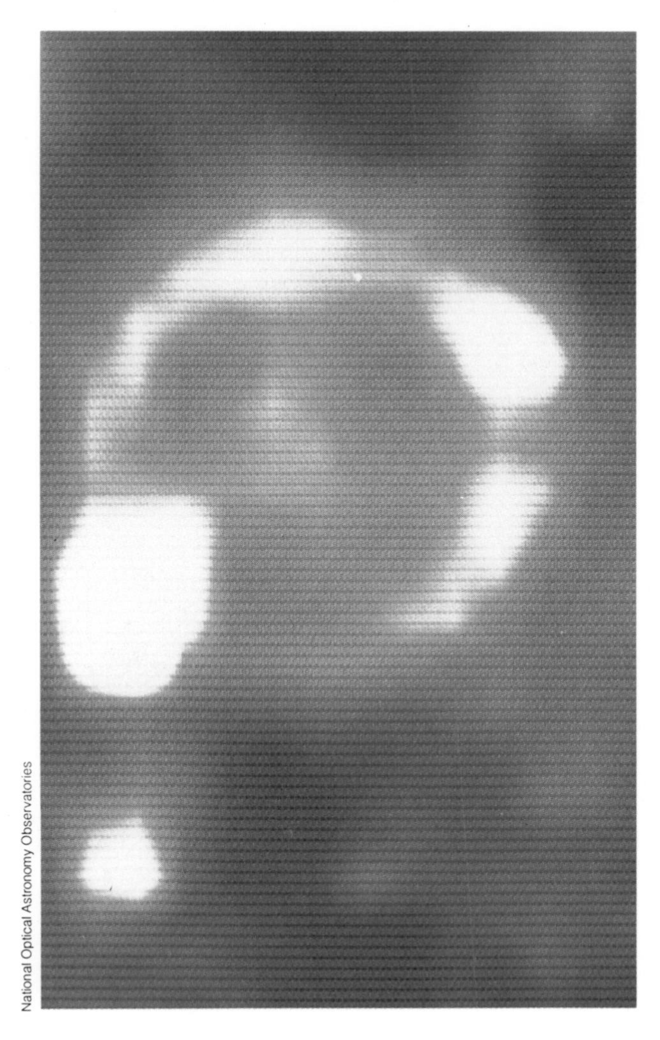

National Optical Astronomy Observatories

The giant dome of the 158-inch (four-meter) Mayall Reflector at the Kitt Peak National Observatory near Tucson, Arizona, dwarfs two nearby domes belonging to the University of Arizona. On the left is the thirty-six-inch (one-meter), and the ninety-inch (2.5-meter) instrument in the center.

National Optical Astronomy Observatories

the infrared universe. They have built large infrared telescopes on high mountaintops where the air is thin and dry. The highest is on Mauna Kea in Hawaii, nearly 14,000 feet (4,267 meters) above sea level—so high, in fact, that visiting astronomers must occasionally take "oxygen-breaks" during the night.

Scientists have even taken their instruments aboard a high-flying airplane known as the *Kuiper Airborne Observatory* (KAO). This modified Lockheed C-141 transport jet carries a thirty-six-inch (1-meter) infrared telescope and a crew of a dozen astronomers to an altitude of 7½ miles (12 kilometers). From here, above 99 percent of the water vapor in our atmosphere, astronomers can study the infrared universe for several hours at a time before returning to Earth.

The Flight of IRAS

On the evening of January 25, 1983, a rocket rode a tower of flame into the clear, dark California sky. It carried with it a revolutionary new telescope. Called the *Infrared Astronomy Satellite*, or IRAS, this telescope was assembled by scientists from the United States, the Netherlands, and the United Kingdom.

Orbiting 650 miles (1,040 kilometers) above the Earth's surface, its twenty-three-inch (0.5-meter) telescope was cooled by liquid helium to less than ten degrees above absolute zero to prevent its own heat emissions from blinding itself.

During its ten month mission, IRAS surveyed the sky twice, and found a quarter of a million infrared sources—more than a hundred times more than were previously known. It discovered six new comets, several asteroids, three dust shells around the giant star Betelgeuse, and strange infrared "cirrus" clouds all around the sky.

Perhaps the most stirring moment of the IRAS mission came when scientists from the Jet Propulsion Laboratory in Pasadena announced their discovery of a dusty disk orbiting the bright star Vega and suggested it may be another solar system in the process of forming.

Amazing New Eyes

It has been said that everything we know about the universe has come from light. Only recently has that begun to change. Today, we know that our universe speaks to us in many different ways: from the mysterious static of the radio, to the ghostly glow of the infrared, across the beautiful visible spectrum of colors, and into the powerful ultraviolet x-rays and gamma rays.

Each is like a piece of the cosmic jigsaw puzzle. For thousands of years we've tried to assemble it with pieces of only certain colors. Today, astronomers have collected many more, and continue their search for others. Only by having them all can the star hunters ever hope to decipher the puzzle we know as the universe.

© Greg Vaughn/Tom Stack & Associates

Courtesy NASA

Above, top: ***A large infrared telescope constructed by the University of Hawaii and NASA sits atop Mauna Kea in Hawaii, far above most of the infrared-absorbing water vapor in our atmosphere.*** **Above, bottom:** ***The Kuiper Airborne Observatory (KAO) is operated by NASA's Ames Research Center in Mountain View, California.***

chapter nine

TO WORLDS BEYOND

On the evening of Sunday, July 20, 1969, more than a billion people huddled around their television sets to watch the most remarkable event in the history of mankind. It was a moment that humans had dreamed of for thousands of years: the landing of a man on the moon.

On the stark lunar plain known as the Sea of Tranquility sat *Eagle*, the Lunar Excursion Module (LEM). Inside, pilot Edwin (Buzz) Aldrin Jr. performed duties on his checklist. Overhead, seventy miles (112 kilometers) above the cratered terrain, Command Module Pilot Michael Collins flew around the moon in *Columbia*, the craft that eventually took the astronauts back home.

But now, with the words "Live from the moon" appearing on TV screens, the world watched as Commander Neil Armstrong slowly descended the ladder from the LEM. He described the surface of the moon to amazed listeners a

On July 20, 1969, astronaut Edwin (Buzz) Aldrin Jr. descended the ladder of the Lunar Excursion Module (LEM), and became only the second man to walk on the moon. This photo was taken by Apollo 11 Commander Neil A. Armstrong.

Courtesy NASA

Courtesy NASA

quarter of a million miles away. As he descended, hearts around the world pounded faster.

And then, at 10:56 P.M., EDT, Armstrong stepped off of the ladder. As his boot touched the powdery soil of our nearest cosmic neighbor, he proudly spoke words that will live forever: "That's one small step for a man, one giant leap for mankind."

Suddenly, the world erupted in joy. People cheered and shouted, and car horns blared. Our age-old dream had become reality. A man was standing on the moon!

Older adults stared in amazement and relived memories of Buck Rogers and Flash Gordon. Youngsters watched with excitement and dreamed of making the journey themselves someday.

For one epic moment, the whole of humanity was one. There were no Soviets or Americans, Jews or Arabs, blacks or whites. For the first time in history, we had become the "people of planet Earth."

On the moon it was silent. Armstrong gazed upward into the black lunar sky at the sparkling blue and white globe he knew as home. This was the world where humanity had evolved and learned through the millenia; where wars were fought, and lives were lost over property, money, and ideals. And it was the world where star hunters of ages past gazed skyward and dreamed of traveling among the stars.

Nearly a century ago, Russian space pioneer Konstantin E. Tsiolkovsky recognized our need to journey to worlds beyond. "The Earth is the cradle of mankind," he said, "but one cannot remain in the cradle forever."

It was on a warm July night in 1969 that we took our first step from the cradle.

THE SPACE AGE BEGINS

The dream of space travel was barely a century old. It was Tsiolkovsky who, in the late 1800s, first proposed using liquid hydrogen and oxygen as fuel for multi-stage rockets, but it was Robert Goddard of Massachusetts who first tried it out.

On May 16, 1926, Goddard launched the first liquid fueled rocket to a staggering height of forty-six feet (fourteen meters). By the 1930s, he was using gyro-controlled rockets with automatic steering devices that could go as high as two miles (three kilometers).

Opposite page: ***Astronaut Aldrin stands on the lunar surface during his* Apollo 11 *Extravehicular Activity (EVA). Since there's no erosion on the moon, the footprints of Armstrong and Aldrin will remain in the dusty soil for millions of years.*** **Left:** ***Floating in orbit above the moon, the Lunar Excursion Module ascent was photographed by Command Module Pilot Michael Collins.*** **Below:** ***On August 26, 1937, Massachusetts rocket expert Robert Goddard launched one of his rockets from Roswell, New Mexico.***

Courtesy NASA

Sovfoto/Eastfoto

Through the years, rockets became more complex, and were launched to even greater heights. But it wasn't until October 4, 1957, that the dream of space travel became a reality.

On that day, the world was stunned when the Soviet Union announced that the first artificial satellite, *Sputnik 1*, had been successfully launched into orbit around the Earth. One month later, the Soviets launched another satellite—*Sputnik 2*. This was no ordinary robot craft, however, for *Sputnik 2* contained a live dog named Laika.

The launches of *Sputnik 1* and *2* were a technological embarrassment to the United States Government, since it had tried unsuccessfully for many years to launch its own Earth-orbiting satellites.

Then, under mounting pressure, the United States hired German rocket scientist Werner Von Braun to direct its effort. Success was not far behind. On February 1, 1958, an Army Redstone booster lifted a craft named *Explorer 1* into orbit. The United States had become a space-faring nation.

Above: Sputnik 2 ***carried a dog named Laika into orbit on November 3, 1957. With this mission, the Soviet Union proved that a living creature could survive a trip into space with no ill effects.*** **Right:** ***The*** **Vostok** ***was the first craft to successfully carry a man in an orbit of the Earth.*** **Below:** ***Yuri Gagarin (1934–1968) became the first man in space when he was launched into orbit around the Earth in*** **Vostok 1** ***on April 12, 1961. He died seven years later during a test flight of a jet aircraft.***

Sovfoto/Eastfoto

MAN IN SPACE

While engineers and technicians at Cape Canaveral prepared for America's first manned orbital flight, identical efforts were being made on the other side of the globe.

Sovfoto/Eastfoto

On April 12, 1961, Soviet Air Force pilot Yuri Gagarin became the first human to journey into space. He successfully piloted his *Vostok* craft through one orbit of the Earth before landing on Soviet soil 108 minutes later. The Soviet Union again was first, but only three weeks later Alan Shepard became the first American to fly in space aboard the craft *Mercury 1.*

The die had been cast. Our eternal bonds of gravity had been broken and space flight had become a reality. On May 25, 1961, with the world eagerly watching, President John F. Kennedy set a goal for the United States: "I believe that this nation should commit itself to achieving the goal, before this decade is out, of landing a man on the moon, and returning him safely to the Earth."

The Soviets had the same goal and the space race was on.

Courtesy NASA

Courtesy NASA

Above: ***While practicing rendezvous and docking maneuvers in December 1965, the*** **Gemini 7** ***spacecraft was photographed through the hatch windows of the*** **Gemini 6** ***spacecraft. The Earth appears 160 miles (257 kilometers) below.*** **Left:** ***In his historic message of May 25, 1961, President John F. Kennedy set a goal of landing a human on the moon and returning him safely to the Earth by the end of the 1960s. Behind Kennedy is Vice President Lyndon B. Johnson.***

Courtesy NASA

The massive* Saturn 5 *rocket lifts off of pad 39A at 9:32*A.M. *EDT on July 16, 1969. Three* Apollo 11 *astronauts rode its first-stage thrust of 7.5 million pounds (3.4 million kilograms) on their way to the moon.

RACE FOR SPACE

During the next seven years, the world watched as both the United States and the Soviet Union tested the hardware and techniques necessary to send humans safely to the moon. The Soviets continued with a series of space "firsts:" the first dual spacecraft, the first woman in space, the first three-man crew, and the first space walk. But it was the United States, always seemingly one step behind the Soviet Union, who reached the moon first.

In December 1968, a *Saturn 5* rocket roared off its Florida launch pad and carried with it the first humans to circle the moon. While in lunar orbit, astronauts Frank Borman, James A. Lovell Jr., and William A. Anders of *Apollo 8* became the first men to watch their home planet rise above the horizon of an alien world.

Then, on July 20, 1969, the spidery LEM of *Apollo 11* touched down on Tranquility Base, and history was made. Kennedy's national goal of landing a man on the moon before the end of the decade was realized with only five months and eleven days to spare.

Because of a failing economy and expensive military involvement in Vietnam, the lunar program was terminated in 1972, and the moon fell silent once again.

To date, twelve men have walked the soil of our nearest cosmic neighbor. They crossed some sixty miles (ninety-six kilometers) of lunar terrain, set up scientific experiments, and returned to Earth with 842 pounds (383 kilograms) of lunar samples. And they inspired an entire generation to reach for the stars.

Courtesy NASA

Courtesy NASA

Courtesy NASA

Left: Apollo 11 *astronaut Edwin Aldrin Jr. takes a lunar soil sample for laboratory analysis on Earth. This process was part of a two hour, twenty minute excursion about the lunar surface to prepare a number of scientific experiments.* **Middle, left:** *Standing on the rim of the thirty-three-foot (ten-meter) deep Plum Crater at the Descartes landing site is* **Apollo 16** *astronaut Charles M. Duke Jr. Parked in the distance is the Lunar Roving Vehicle with which the astronauts toured the lunar terrain.* **Middle, right:** *Edwin Aldrin walks away from Neil Armstrong as he prepares to deploy two scientific experiments on the lunar surface.* **Below:** *In an effort to prevent the possible spread of viruses that may have been picked up on the moon, Lt. Clancey Hatleberg scrubs down the returning* **Apollo 11** *astronauts with disinfectant. Nearby, their capsule floats in the waters of the Pacific Ocean.*

Courtesy NASA

Right: ***A view of Mars.*** **Below:** ***A thermal test model of the Viking Orbiter spacecraft is being tested in a thermal-vacuum chamber at the Jet Propulsion Laboratory in Pasadena, California. Its television cameras are mounted on the scan platform at lower left.***

Courtesy NASA

Courtesy NASA

RENDEZVOUS WITH MARS

The moon was not the only target of spacecraft. Robot vehicles had been launched toward several nearby planets as well, and returned spectacular photos to their creators on Earth. Perhaps most stunning of all was Mars.

For most of human history, Mars appeared as a silent, orange light that moved mysteriously among the stars from night to night. But when Percival Lowell first aimed his large telescope in its direction in the late 1800s, ideas began to change.

Astronomers of the time viewed Mars as a world of lush vegetation that changed with the Martian seasons—from the greens of spring to the browns of fall. Some believed that this world was populated by a race of intelligent beings who constructed huge canals to irrigate their crops with water from melting polar ices.

Not until the 1960s did the technology to explore Mars catch up with our age-old desire to do so. In 1964, only thirty-three days from year's end, a rocket lifted off from the eastern coast of Florida, and carried with it a robot spacecraft on a history-making flight to Mars. Its name was *Mariner 4*.

Seven months later the craft arrived at its destination. On July 14, 1965, at 5:18 and 33 seconds P.M. PDT, the world watched as *Mariner 4* snapped the first closeup picture of Mars.

The probe shot twenty-two pictures in all. They showed an ancient, dead world—a bleak, 3½-billion-year-old, cratered terrain. This was not the Mars of myth. Where were the canals? The lush fields of vegetation? And where were the Martians? Suddenly, the Mars of our imaginations gave way to a Mars of mystery.

During the next few years, other spacecraft visited Mars, and all showed the same cratered land. But one, *Mariner 9*, changed forever our view of the Red Planet. Its mission was to orbit the planet and photograph more of the Martian terrain than its predecessors.

When it arrived at Mars in 1972, *Mariner 9* radioed more than 7,300 photos to Earth, re-

Courtesy NASA

An Atlas-Centaur rocket rises from its Florida launch pad, carrying with it the* Mariner 7 *spacecraft bound for Mars. It was the second half of the two-spacecraft* Mariner *mission to the red planet in 1969.

vealing wonders no human eyes had ever seen. There were immense, extinct volcanoes, one towering nearly sixteen miles (twenty-five kilometers) above the surrounding terrain; a canyon the length of North America, forty-three miles (sixty-nine kilometers) wide and five miles (eight kilometers) deep; and a global network of dried-up gullies carved by once-raging rivers in Mars' distant past.

Surely a world so bleak and dry couldn't be home to life. Or could it? This was the very question that scientists hoped to answer when they launched the twin robot crafts *Viking 1* and *2* to Mars.

Is There Life?

On July 20, 1976, just seven years after humans first landed on the moon, *Viking 1* set down on Mars. It landed on the flat, dusty plains of Chryse Planitia. Two months later, it landed on the opposite side of the planet.

Their cameras showed rolling hills, rocks, boulders, orange soil, and a strange pink sky. If any life existed here, their cameras didn't show it. Their mechanical arms dug into the Martian ground and analyzed the soil for microorganisms. The results were inconclusive.

Meanwhile, high overhead, the *Viking* orbiters photographed nearly every square mile of the Martian surface with unprecedented clarity. They showed no signs of canals or canal builders. They found no footprints, no obvious artifacts, no trees, bushes, or animals; not even as much as a complex organic molecule beneath the rusty soil.

Finally, on May 21, 1983, radio contact with *Viking* was lost. And Mars, which had only begun to reveal its innermost secrets, fell silent once again.

Voyage to the Outer Planets

In the meantime, spacecraft from the United States and the Soviet Union had visited Mercury, Venus, Jupiter, and Saturn, and had radioed back closeup photos of each. But perhaps the most exciting of all, was a mission known as Voyager.

In August and September of 1977, two Titan rockets lifted off from Cape Canaveral, Florida, carrying two robot craft on a journey that would cover billions of miles. *Voyagers 1* and *2* were headed toward the outer planets of our solar system.

Courtesy NASA

The first photo returned from the surface of Mars showed a field of red soil and rocks. The sky appeared a strange pink color from the red dust that is blown around by Mars' tremendous winds. The photo was taken by the* Viking 1 *robot lander in 1976.

Left: ***Jupiter appears in this*** **Voyager 1** ***photo taken in 1979 from a distance of only 25 million miles (40 million kilometers). One of Jupiter's large Galilean moons, Ganymede, can be seen in the lower left of the photograph.*** **Below:** ***In preparation for launch, a*** **Voyager** ***spacecraft model is used for electrical and countdown tests at the Kennedy Space Center in Florida.***

Encounter with a Giant

In 1979, the *Voyagers* first encountered the giant planet Jupiter, and began to send back thousands of images and measurements of the most violent storms in existence.

At this world, the spacecraft photographed colorful clouds, twisted and stretched into a myriad of shapes creating glorious objects of art as well as wonders of science. We saw cyclones larger than Earth tear through the clouds at hundreds of miles per hour. The largest, Jupiter's Great Red Spot, has raged continuously for at least 400 years. In its thick, turbulent atmosphere, we saw lightning bolts longer than continents on Earth.

As the probes sped by the giant planet, the *Voyagers* photographed its many moons. Scientists on the ground learned that most, like Europa and Ganymede, were ice-covered worlds—just as they had expected.

Right: ***The Great Red Spot of Jupiter, a cyclone some three times larger than the Earth, swirls in the atmosphere of Jupiter. The storm has been raging for at least three centuries.*** **Below:** ***The most volcanically active moon of the solar system is Jupiter's moon, Io. An enormous volcanic plume can be seen erupting against the blackness of space. Io erupts enough material in one year to completely recoat its surface.***

Courtesy NASA

Courtesy NASA

As scientists watched the edge of the moon Io, they got a surprise. While looking for a faint star near the edge of Io, engineer Linda Morabito exaggerated the brightness of the image on her computer screen. When she did, she saw a great umbrella-shaped plume on its edge. "I'd never seen anything like this before," she recalled, "and I suspected that no one else had either."

It was an erupting volcano, the first ever found on a world other than Earth. It rose 174 miles (278 kilometers) above the Ionian surface, or thirty times higher than Mount Everest. It wasn't the only one, however, for when they examined the moon closely, scientists found at least six more volcanoes blasting material into empty space.

As the *Voyagers* swung around Jupiter and left it behind, they looked back at the planet and photographed the thin dark rings that encircle it. Then they set their courses toward their next encounter some two years, and 2½ billion miles (four billion kilometers) away: Saturn.

Courtesy NASA

Above: ***From a distance of only 8.6 million miles (13.9 million kilometers),*** **Voyager 2** ***photographed the ringed planet Saturn with greater clarity than ever before.*** **Inset:** ***In this artist's conception, the*** **Voyager** ***spacecraft flies past Saturn and analyzes the makeup of its ring system.***

Visit to Saturn

In 1981, scientists on Earth sent the two craft a "wake-up call" to prepare them for arrival at Saturn. When *Voyager I* arrived, it found the atmospheric storms there to be fewer and more subdued than on Jupiter, but the winds were three times more powerful, tearing along at nearly 1,100 miles (1,760 kilometers) per hour. Most spectacular were the planet's magnificent rings. Within the rings themselves, the cameras of *Voyager 1* revealed undreamed-of complexities.

As *Voyager 1* swung around the back side of the planet, it was sent off into deep space with other work to accomplish. By now its sister craft, *Voyager 2,* had arrived at Saturn and began photographing the icy moons. Just then, its camera mounting jammed and began sending back to Earth only photos of black space. And then, as the craft flew through the ring plane on its way out of the Saturnian system, it was struck by a chunk of ice.

***An artist's concept of Uranus and its rings. This painting was taken from images returned on January 17, 1986 from* Voyager 2.**

Meeting with the Green Giant

Its delicate magnetometer arm began to wobble dangerously. Scientists on Earth were concerned that they might not be able to recover the spacecraft, and that its mission might be over.

Soon, the craft was out of danger as it left the Saturnian system behind. It was headed outward toward its next encounter with the mysterious blue-green planet: Uranus.

Engineers pondered its problems, for if they couldn't be fixed, the craft would sail blindly by the planet, losing its only chance to photograph this mysterious world.

It took four and a half years, but engineers on Earth solved the problems of *Voyager 2*. They taught it new ways to gather data and transmit it to its rapidly disappearing home. And, as the craft closed in on Uranus in early 1986, all eyes were on the mission.

Since William Herschel first spotted Uranus in 1781, it had appeared as merely a tiny green dot in even the largest of telescopes. Now, scientists hoped to get their first closeup look at this world.

On January 24, 1986, *Voyager 2* whipped through the Uranian system, snapping thousands of pictures as it went. It showed a quiet atmosphere compared to those of Jupiter and Saturn. It saw a thin, icy fog high in the planet's cold atmosphere, and a strange electro-glow coming from its sunlit side.

On its way by, *Voyager 2* photographed the thin dark rings encircling Uranus, and the mysterious satellites seen only as tiny dots in Earth-bound telescopes. One of the most unusual was Miranda.

Photographs of this world showed a peculiar landscape that betrayed a violent past. One of the most mind-boggling features was a treacherous cliff twelve miles (nineteen kilometers) high.

Its job at Uranus complete, *Voyager 2* continued to operate remarkably well and scientists on Earth sent it on its way toward its next encounter. Another long, lonely journey lay ahead, this time three and a half years to its final destination: Neptune.

Distant Neptune

In the spring of 1989, *Voyager 2* began to photograph Neptune. Now three billion miles (five billion kilometers) from the warmth of the sun, *Voyager*'s internal heaters were working almost constantly; its memory was fading fast. It had become partially tone-deaf, and was slowly losing power in its three nuclear generators. Radio signals sent to Earth were now taking four hours to arrive.

And then, on August 24, 1989, *Voyager 2* streaked through the Neptunian system at 61,000 miles (97,600 kilometers) an hour and photographed everything in sight for transmission back to Earth.

It showed a world of almost translucent blue color, streaked with thin, white methane clouds. Here we saw a great swirling storm the size of Mars, and smaller storms the size of continents on Earth.

And at Neptune's largest moon, Triton, *Voyager* photographed a world of nitrogen and methane ices. Here was a world where temperatures hover at 400 degrees Fahrenheit (204 degrees Centigrade) below zero, and where surface features were so odd that they have not yet been explained. During its flyby, *Voyager 2* photographed two nitrogen geysers, which spew dark material onto the moon's icy surface.

A Remarkable Mission

More than a dozen years, millions of bits of data, and nearly a hundred thousand photographs later, *Voyager 2* continues to astound scientists.

The size of a subcompact car, each *Voyager* craft contains some five million working parts. They sent data and photographs back to waiting scientists on Earth billions of miles away by means of a transmitter with a power of only 23 watts (about the same as the light bulb in your refrigerator!) After twelve years, and a journey of more than 4½ billion miles (seven billion kilometers), *Voyager* skimmed along the cloud tops of Neptune within only twenty miles (thirty-two kilometers) of its target point, and did so within one second of its schedule!

To the Stars

Their tour of the planets is complete, but their mission is not, for both *Voyagers* will spend the next decade or two searching for the boundary between the atmosphere of our sun and the space between the stars. And when they find it, they will leave behind their planetary family forever, and drift outward toward the stars.

One day, two or three decades from now, their power will weaken, and scientists on Earth will lose contact with them. But even in death, their mission will continue, for on their sides, each carries a golden phonograph record with the sights and sounds of the Earth and its people.

Someday, perhaps thousands or millions of years from now, the *Voyagers* may encounter other beings who share the cosmos with us. And from their disks, the discoverers may learn of the creative and inquisitive beings who long ago sent the robots on an exciting and spectacular voyage to worlds beyond.

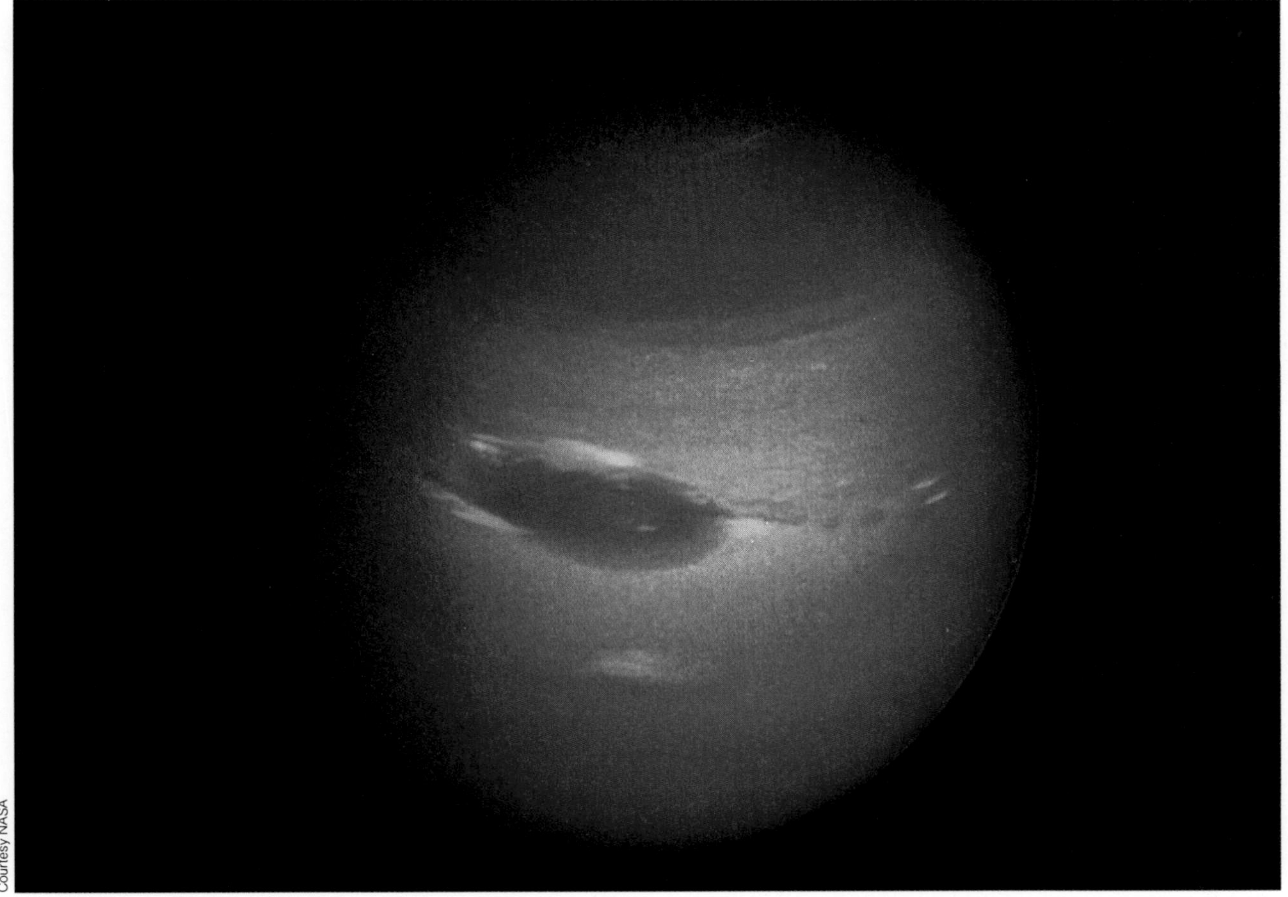

Courtesy NASA

This picture of Neptune was taken by* Voyager 2. *It clearly shows the details of the cloud structure. The Great Dark Spot and the high southern latitudes have a deep blue cast.

NASA
USA
NASA

chapter ten

STAR HUNTERS OF TOMORROW

From an Earth-orbiting space station, humans will be able to perform research and create products impossible on Earth. It might be used to build larger space equipment or even serve as a transfer point for longer manned missions to the moon or Mars.

Courtesy NASA

The seatbelt sign is flashing, and the captain reminds us to take our seats. Through the window we see sunlight glistening off the magnificent blue and white globe below. Overhead, against the pitch black sky, a huge, white structure seems to grow ever larger as we approach. Floating nearby are dozens of shuttles, some coming and some going. They appear so tiny that they look like gnats around a barn.

Soon, we feel a thud. We are docked. The passengers gather their belongings and go their separate ways. Some are astronomers heading for work on Lunar Colony 5. Others are Mars shuttle pilots arriving to relieve their colleagues of duty, repair crews heading toward orbital gravity-wave telescopes, and engineers on their way to the mining camp at the L2 point behind the moon.

At the departure gate, a lunar girl scout troop anxiously

USA
NASA
Discovery

awaits its first flight to Earth. And down the long white corridor, United Planets delegates board a charter craft to the transfer station on Deimos. For several months they have been visiting Earth to discuss mineral trading rights and are now heading home for the holidays.

A scene from the latest science fiction film? No. A page out of the future of humanity? Quite possibly.

We now stand on the threshold of an exciting new era in the study of our universe. We are developing technology to visit the planets, asteroids, and comets of our solar system. In addition, we are launching telescopes into Earth-orbit to view the many radiations the universe creates, and are building larger and larger telescopes on the ground.

The star hunters of tomorrow are in for a most exciting time. But their work begins today, right here on planet Earth.

Courtesy NASA

GIANT EYES OF TODAY

Until recently, the largest telescope on Earth was the 200-inch (five-meter) reflector on Palomar Mountain, in California. With a weight of some 500 tons (454 metric tons), this giant was long believed to represent the upper limit of large ground-based telescopes. In fact, for more than a quarter century, no one even tried to make a larger one.

And then, in 1976, an even larger telescope was erected in Zelenchukskaya, U.S.S.R. It was 236 inches (six meters) across, but suffered many structural problems, including an inability to produce sharp images. To many astronomers, these problems confirmed their suspicions that the upper limit of telescope size had been reached.

Others weren't so sure. They believed that larger telescopes were possible, but to make them, astronomers had to invent totally new ideas of what telescopes "should" look like.

Scientists began to try unusual designs. They built telescopes not from one large mirror, but from several smaller ones coupled together. They've created large, lightweight, single mirrors by spinning molten glass while it was cooling to give it a curved surface. And they've developed extremely thin mirrors whose precise shapes are controlled by computers.

The first of these revolutionary instruments was built atop Mount Hopkins in southern Arizona in 1979. The Multiple Mirror Telescope (MMT) used six, seventy-two-inch (1.8-meter) mirrors mounted on a common structure to gather as much light as a single 178-inch- (4.5-meter-) diameter instrument.

Its mounting—a simple altitude-azimuth mount—is driven by a computer to follow the stars on their nightly jaunts across the sky. And, instead of a familiar dome, the MMT is enclosed by a barn-shaped housing that reduces air turbulence and helps produce some of the sharpest images possible.

At a fraction of the size and cost of a traditional 178-inch (4.5-meter) instrument, the MMT set the stage for a totally new generation of twenty-first century telescopes.

Opposite page: ***The space shuttle*** **Discovery** ***stands on its launch pad at Cape Canaveral awaiting its 1989 launch. It was the first launch of a shuttle since the tragic*** **Challenger** ***accident three years earlier, and marked America's long-awaited return to space exploration.*** **Above:** ***The space shuttle will provide a vital link between Earth and future space stations, carrying crews and equipment back and forth. In this artist's conception, a shuttle is docked with a pressurized resource node to allow crew members to transfer safely to the station.*** **Below:** ***On a future lunar base, scientists and engineers will study the moon's interior in greater depth. In this artist's conception, a deep-drill team obtains core samples from the floor of the large crater, Aristarchus.***

Courtesy NASA

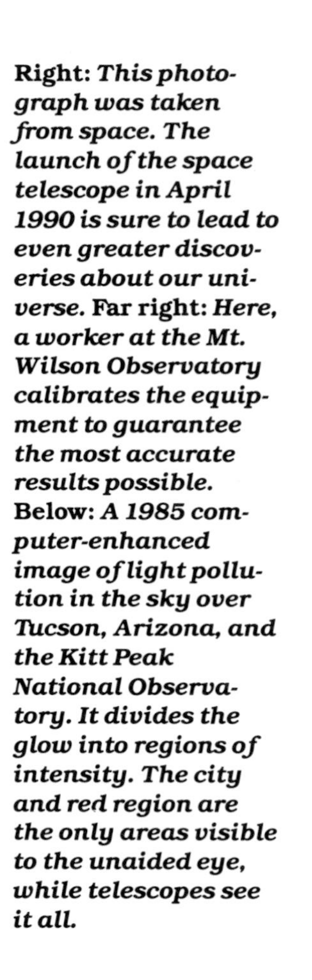

Right: ***This photograph was taken from space. The launch of the space telescope in April 1990 is sure to lead to even greater discoveries about our universe.*** **Far right:** ***Here, a worker at the Mt. Wilson Observatory calibrates the equipment to guarantee the most accurate results possible.*** **Below:** ***A 1985 computer-enhanced image of light pollution in the sky over Tucson, Arizona, and the Kitt Peak National Observatory. It divides the glow into regions of intensity. The city and red region are the only areas visible to the unaided eye, while telescopes see it all.***

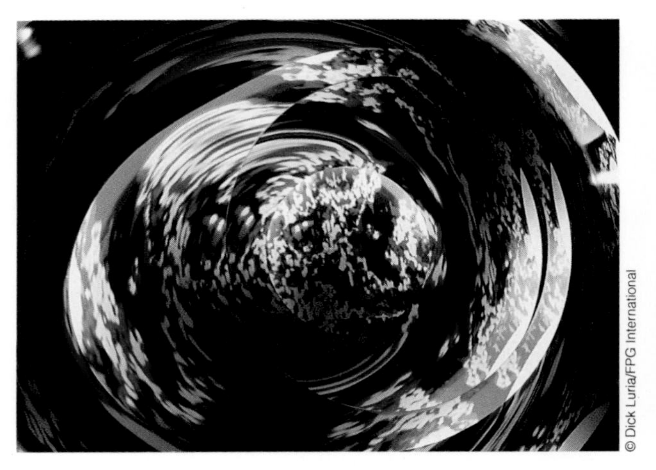

© Dick Luria/FPG International

© Dan McCoy/Rainbow

GIANT EYES OF TOMORROW

In 1991, the Keck Telescope will open its eye on the universe. Now being built on Mauna Kea by scientists at the California Institute of Technology and the University of California, its mirror will be made of thirty-six, seventy-two-inch (1.8-meter) segments fit together like bathroom tiles into one giant eye, 400 inches (ten meters) across. Each segment's position will be monitored by sensors and a computer to keep them aligned to an accuracy of .000001 of an inch (.0000025 of a centimeter).

By the mid-1990s, the European Southern Observatory hopes to open an array of four, 315-inch (eight-meter) telescopes on a mountain top in Chile. Each telescope would be enclosed by a collapsible dome, and when coupled together would produce the light gathering power of a single 630-inch (sixteen-meter) telescope—ten times greater than that of the Palomar giant.

OTHER NEW TECHNOLOGY

Such monster eyes will not be working alone. All the optical giants of the twenty-first century will be coupled with revolutionary electronic and computer equipment to utilize their full potential.

Many of the great discoveries of the future will be made not on photographic plates, but

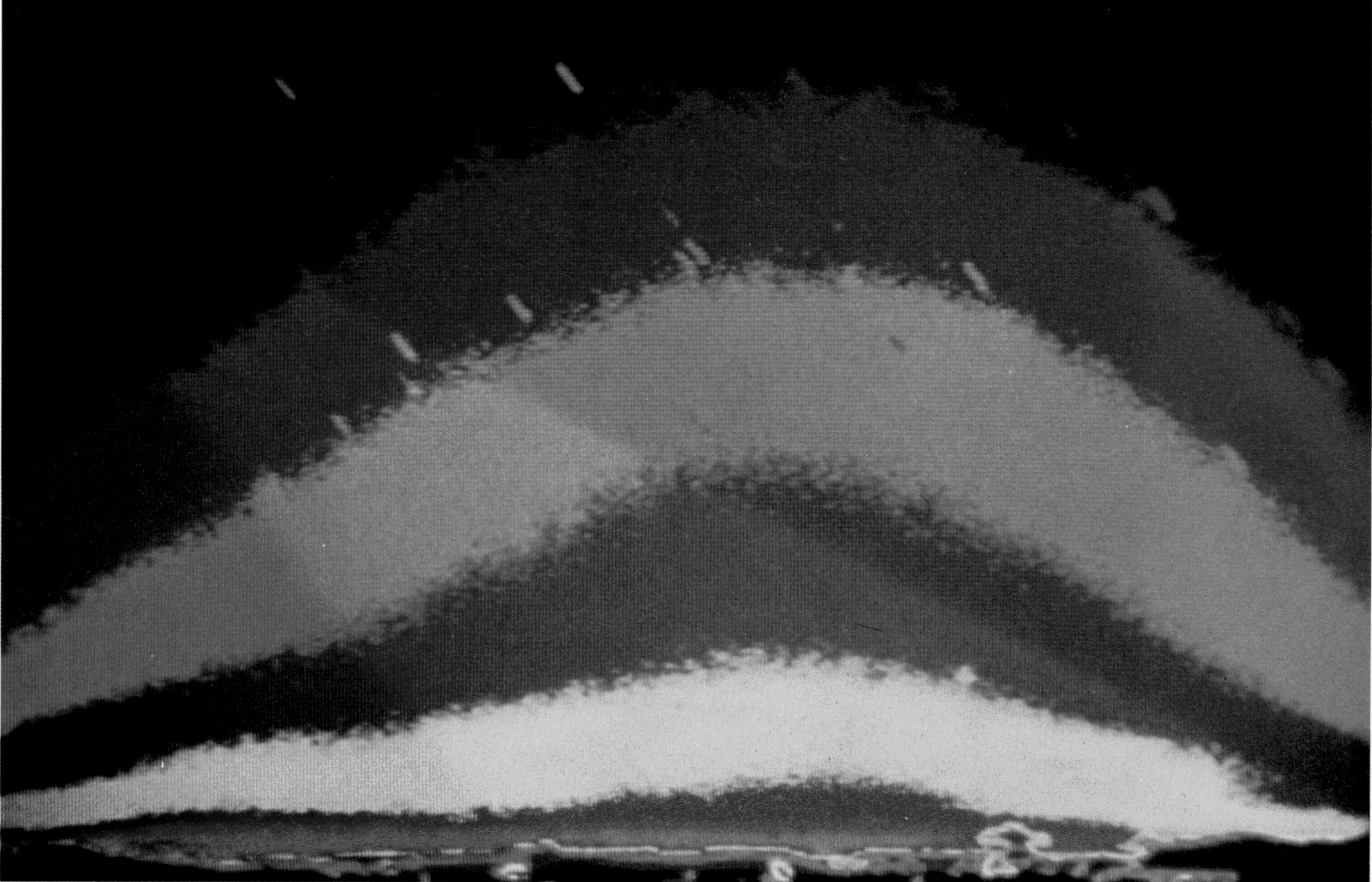

National Optical Astronomy Observatories

on high-tech electronic detectors being developed and tested today. One such detector, called the Charge Coupled Device (CCD), has already begun to revolutionize astronomy.

A silicon chip hardly larger than a dime, the CCD is placed at the focus of a telescope to record light and electronically create an image. A CCD gathers light thousands of times more efficiently than even the best photographic plates, making it possible for an astronomer to capture in only minutes light that once required hours or even days to photograph. In addition, the astronomer can use the electronic information to enhance particular features of the image, or to create many different photographs—all from just one exposure.

Scientists are also developing techniques to produce crystal-clear images of celestial bodies free from the distortions caused by our turbulent atmosphere. When applied to the 200-inch (five-meter) Hale reflector, these techniques have improved the telescope's ability to see detail by twenty times. Astronomers are anxiously awaiting using it on the Keck telescope during the next decade, for they expect to be able to see detail as small as a dime viewed from nearly 6,000 miles (9,600 kilometers) away.

Improvements in computer and communications systems are enabling astronomers to experiment with a process called "remote observing." Future star hunters may not even need to be at the observatory to carry out their research. They may instead sit in their offices or living rooms elsewhere in the world, control the telescope by telephone and computer, and watch their program objects on television screens.

THE SPACE TELESCOPE

As exciting as the future of ground-based astronomy appears, its success is limited by a more mundane and practical problem called light pollution. Today, few places remain on Earth where astronomers can see the sky as our ancestors saw it. Outdoor lighting is needed at night by cities for safety, but much of this light is unshielded and scatters upward where it washes out the sky.

As cities continue to grow, so too does light pollution, and astronomers are quite concerned. The famous Mount Wilson Observatory has been rendered nearly useless by the phenomenal spread of light pollution in Los Angeles. Even more remote sites like the Kitt Peak National Observatory high above the

From Kitt Peak, the sky over Tucson glows brightly and greatly hinders observations of the faintest, most distant objects in the universe. Many cities like Tucson have enacted ordinances that restrict the types of nighttime lighting that can be used. Overleaf: The Mt. Wilson Observatory.

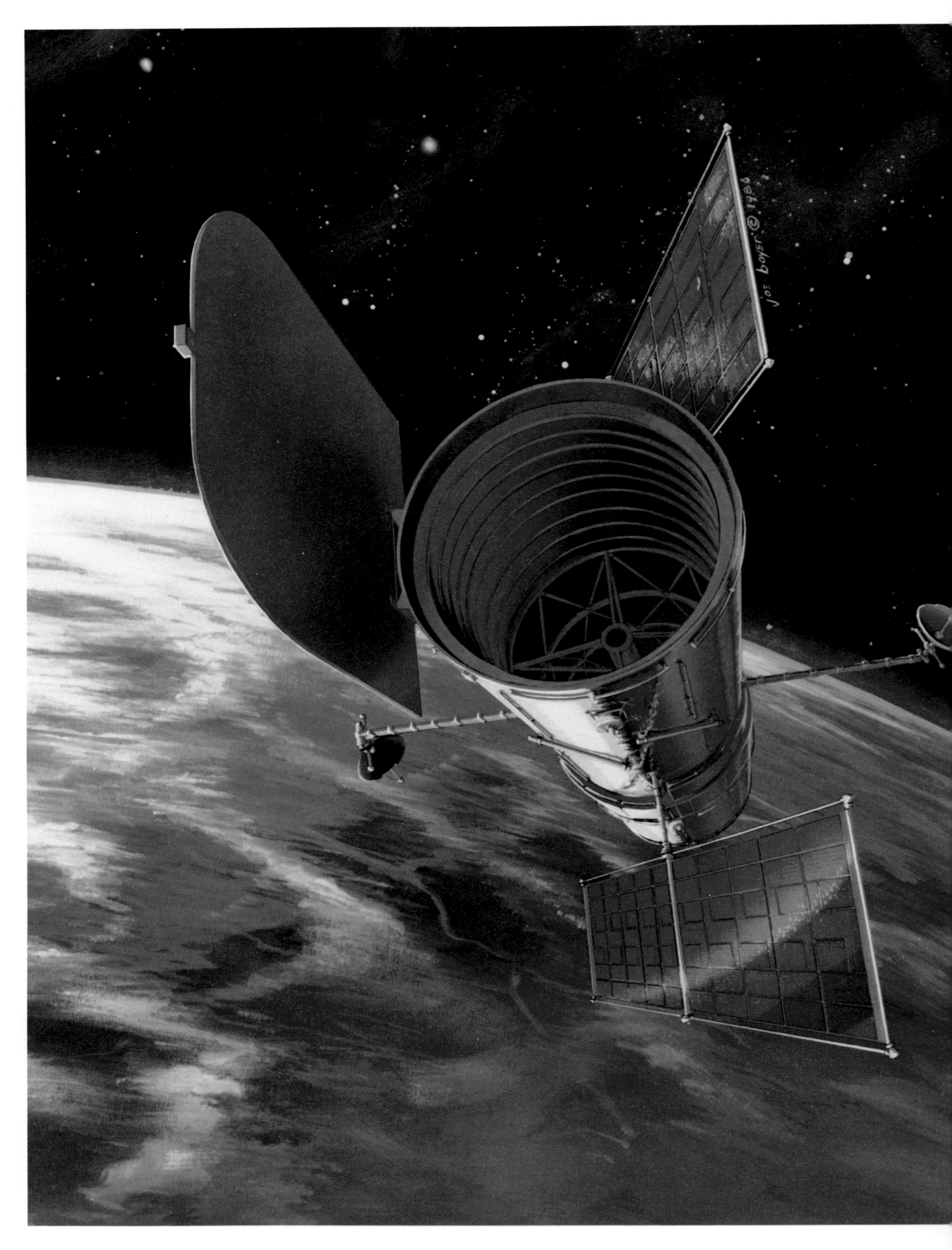
Joe boyer © 1986

Courtesy NASA

desert of Southern Arizona are not immune to the problem caused by growing cities.

Another problem is the atmosphere under which we live. Weather, turbulence, and daylight limit the valuable time with which astronomical discoveries might be made. In addition, our air blocks many of the electromagnetic radiations which carry important clues to the nature of the universe.

For these reasons, scientists are developing plans to launch a series of orbital telescopes to study the universe. The first in the series was launched in the spring of 1990 by the United

Courtesy NASA

States Space Shuttle *Discovery*. Its name is the Hubble Space Telescope.

Named for the famous American astronomer Edwin P. Hubble, the HST contains a 94½-inch (2.4-meter) mirror, polished to an accuracy of half a millionth of an inch. On board, six instruments and cameras will combine to make this the largest and most complex space observatory ever.

Once in space, the Shuttle payload bay doors opened, and the telescope was deployed into its 300-mile (480-kilometer) high orbit to begin its scrutiny of the heavens. During its fifteen-year lifetime, the HST will take advantage of the clear, dark skies of space and will provide astronomers with images ten times sharper than even the largest ground-based telescopes.

Above: ***This artist's rendering of the space shuttle orbiter demonstrates how the space telescope was deployed into space.*** **Left:** ***The Hubble Space Telescope orbits some 380 miles (612 kilometers) above the obscuring atmosphere of the Earth. The telescope was deployed in April 1990, and represents one of a family of observatories being launched to examine the electromagnetic spectrum from end to end.***

THE GREAT SPACE OBSERVATORIES

The HST marks humankind's first permanent astronomical research effort from space. Early in the next century, an entire family of space

Courtesy NASA

Above: ***In 1989, the*** **Magellan** ***spacecraft left the Earth on a fifteen month journey to the cloud-covered surface of Venus.*** **Magellan** ***marked the first United States planetary mission launched in a decade, and the first ever to be deployed by the space shuttle.*** **Right:** ***United States and Dutch technicians prepare the Infrared Astronomy Satellite (IRAS) for launch. IRAS spent nearly a year surveying the entire sky with its infrared eyes, and produced the first systematic study of the heavens in this invisible radiation.***

observatories may be in operation, scanning the electromagnetic spectrum from end to end. What discoveries they will make, no one can imagine.

Through the Gamma Ray Observatory (GRO), astronomers hope to be able to probe the most violent stellar and galactic phenomena, including the elusive and enigmatic black holes.

X-rays will be studied through the Advanced X-Ray Astrophysics Facility (AXAF). A hundred times more powerful than any x-ray telescope ever used, AXAF will study dying stars and the formation and evolution of the entire universe.

The Space Infrared Telescope Facility (SIRTF) will be cooled to nearly absolute zero, and will detect and study in detail the faint wisps of cosmic heat a thousand times fainter than those seen by IRAS. From its 558-mile- (893-kilometer-) high orbit, SIRTF will probe the giant molecular clouds of space to understand the mysteries of stellar birth, and will analyze the properties of cooler bodies throughout our solar system: planets, asteroids, comets, and dust.

Courtesy NASA

Courtesy NASA

The Hubble Space Telescope is being lowered for installation of the sun-shield onto its front end. It is covered with a thermal protective coating.

When coupled with large ground-based optical and radio instruments, these space observatories will give us our first look at phenomena in every radiation possible. They will be many times more sensitive than any telescope now in existence and offer the promise of a major revolution in our understanding of the universe.

BEYOND TOMORROW...

This is only the beginning. Astronomers are developing ideas for giant orbiting space platforms onto which large future telescopes can be mounted. One day, the vast majority of astronomical observations may be made from the clear, dark skies of space.

Astronomers now dream of a Very Large Space Telescope (VLST) to be constructed during the next century. Carried into Earth-orbit by the space shuttle, a giant 315-inch (eight-meter) mirror would be fixed into place inside a tube made from the external fuel tank.

Other more radical designs might also make their way into space. One, called the Coherent Optical System of Modular Imaging Collectors (COSMIC) might use several mirrors in a cylindrical housing to produce a single image much as the MMT has done. COSMIC could be expanded into a huge array of 118 feet (thirty-six meters) across and would see detail 300 times finer than is now possible with the largest telescopes.

But even this huge instrument might soon be dwarfed by a major project known as the Thinned Aperture Telescope (TAT). This instrument would use an enormous ring of mirrors 300 feet across to collect light. Not only would this monster be able to spot planets in orbit around distant stars, it might also enable astronomers to study their surface or atmospheric properties as well!

And someday we may see giant radio dishes electronically linked together in Earth-orbit. They would detect faint whispers of cosmic radio emissions with the clarity of a single dish tens of thousands of miles across. These arrays, too, may someday grow to extend across our entire solar system, producing radio images hundreds of thousands of times sharper than those available today.

Right: ***The*** **Galileo** ***spacecraft with its inertial upper stage is placed into the payload bay of the space shuttle*** **Atlantis. Galileo** ***was deployed in 1989 on a six year journey to Jupiter, to orbit the planet and drop a probe into its atmosphere.*** **Below:** ***From the aft flight deck of the space shuttle*** **Atlantis,** ***astronauts photographed the deployment of the*** **Magellan** ***spacecraft in 1989 as it begins its long journey to Venus.***

Even our moon may one day be developed as an astronomical observing station, with radio and optical telescopes scattered about its far side. Shielded from the light and radio interference of their home planet, star hunters here could use the 325-hour-long lunar night to collect celestial radiation for days and weeks at a time.

And astronomers may place high in Earth-orbit a giant antenna to detect "gravity waves" as they ripple through space. This antenna would contain loosely mounted mirrors on the ends of its arms and would be monitored for any slight movement caused by the passage of a wave. If detected, these waves may lead to our understanding and, perhaps, harnessing of one of nature's most fundamental forces—gravity.

Courtesy NASA

ROBOTS IN SPACE

Our most exciting advances may come from first-hand exploration. We will soon witness a voyage to Jupiter by a robot craft named *Gali-*

Courtesy NASA

The* Mars Observer *orbits the red planet in this artist's conception. It will follow up on the knowledge gained from the* Mariner *and* Viking *missions to Mars during the past three decades.

leo. Launched from the space shuttle *Atlantis* in October 1989, *Galileo's* six-year mission is to orbit the giant planet, and send a probe into its stormy atmosphere.

Also launched in 1989 was *Magellan*. After its encounter with Venus in late 1990, *Magellan* will orbit the planet for 243 days and will map the cloud-covered Venusian surface with radar waves.

NASA is now preparing the *Mars Observer* for launch in the early years of the 1990s. It will orbit the planet Mars in order to measure its climate and chemical makeup, and will photograph the planet like never before. It may even find water beneath the orange soil.

Later this decade, scientists hope to launch the Comet Rendezvous Asteroid Flyby mission, also known as CRAF. CRAF will leave Earth behind for a closeup look at comet Kopff. The spacecraft will fly along with the comet for three years and, in 2001, will deploy the first probe to land on a comet's nucleus.

Later in this decade, a probe named *Cassini* will be launched toward Saturn. *Cassini* will orbit the ringed planet, and a probe will descend into the thick, orange atmosphere of Saturn's largest moon, Titan. Its atmosphere may be similar to the atmosphere of the Earth

A cutaway view of the inside of a material processing laboratory that may eventually be part of a permanently manned space station. Crew members are shown working inside in this artist's conception.

four billion years ago, and may provide scientists with clues to the origin of life. They may even detect the presence of microorganisms thriving in the organic "goo" that may cover the moon's surface.

And by early in the twenty-first century, Mars may have other visitors from planet Earth: balloons which float in the thin Martian air, roving vehicles to venture miles about its terrain, and spacecraft to return to Earth with a sample of Martian soil. And where these probes go, humanity is sure to follow.

JOURNEY TO MARS

We will soon see permanent bases in Earth-orbit and on the moon—settlements in which scientific and industrial work may be performed free of the gravity and atmosphere of Earth.

But beyond the moon, another world beckons: Mars. Early in the twenty-first century, an international crew from Earth may embark on a three year mission to the Red Planet. It would mark our first steps on another planet—as momentous as the voyages of Magellan and Columbus centuries ago. These brave explorers will pave the way for human settlements on this distant world.

In small scientific camps, twenty-first century researchers will study first-hand the age-old mysteries of Mars. Eventually, these Martian settlements will grow, life will flourish, and the umbilical cord from Earth will be cut.

Decades after those first brave explorers set foot on Mars, journeys to the Red Planet may become routine. Passengers will first catch a craft to the Earth-orbiting station where they will shuttle to the Libration Point Spaceport. Here they will catch a ride to the "cycling" ship that runs between Mars and Earth.

On the "cycler," passengers will settle back for the six-month journey. They will experience a rotational gravity similar to that on Earth. But as they near their destination, the gravity will be adjusted downward toward that of Mars.

Once near Mars, they will ride another transfer vehicle to an orbiting spaceport where they can catch a lander going to Mars or its moons.

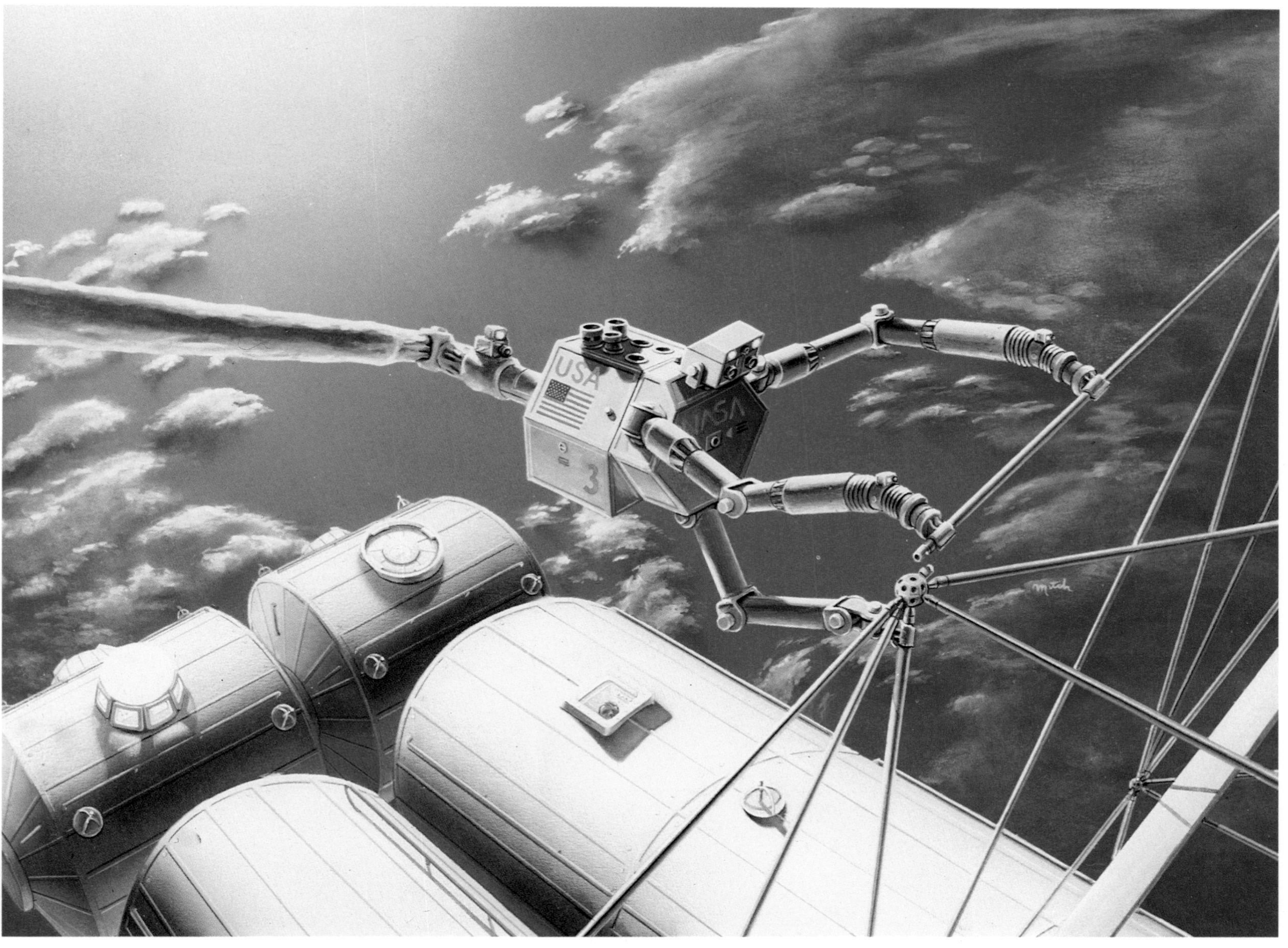

Courtesy NASA

Courtesy NASA

Above: ***The Flight Telerobotic Servicer (FTS) assembles sections of the permanently manned space station's truss structure, as shown in this artist's conception. The FTS will help maintain scientific instruments as well as service visiting spacecraft.*** **Left:** ***In a torus-shaped space colony, farming areas might be interspersed with more populated areas. Artificial gravity might be provided to the 10,000 inhabitants by rotating the colony once every minute.***

Courtesy NASA

Future Mars explorers might visit the caldera of Nix Olympica, the largest volcano in the solar system, seen here in a NASA model. The volcano measures some 375 miles (600 kilometers) across at its base, while its walls tower fifteen miles (twenty-four kilometers) above the surrounding terrain.

Astronauts on Mars will need to be prepared for cold temperatures. This photo of the Martian surface, taken by the Viking 2 lander in 1976, shows a thin frost of water-ice coating the planet's red soil. Temperatures here range from 78°F (26°C) to -168°F (-111°C).

Courtesy NASA

Traveling Biospheres

A round trip to Mars might take between two and three years. No ship could ever carry enough food, water, air, and fuel for such a long journey. So scientists are now experimenting with techniques that might allow such space travelers to "live off the land" during their voyages.

In the desert of southern Arizona, scientists are testing a five million-cubic-foot (4.1 million-cubic-meter) habitat that will be totally closed and self-supporting. Its name is *Biosphere 2 (Biosphere 1* is the Earth).

Within the facility are seven major ecologies: a tropical rainforest, a savannah, a marsh, a salt water ocean, a desert, a small farm, and a human habitat. The geology, weather, and hydrology of each flows smoothly across well-planned boundaries to help control the system.

In late 1990, eight "biospherians" will enter the two-acre habitat and close the doors behind them for two years. During their stay—about the length of a mission to Mars—the biospherians will perform countless experiments to test its life-supporting capabilities without outside influence. And everything around them—from food to air to water—will be recycled by totally natural processes.

A New Mars

During the next century, our descendents may see the landscape of Mars dotted with such biospheres. But one day, perhaps centuries from now, humans may begin the monumental task of returning this harsh, alien world to one like Earth.

Microscopic plants and animals brought there may help Mars replenish its lost oxygen. And giant orbital mirrors may beam sunlight and heat onto its polar ices.

Slowly at first, then more rapidly, the planet would thaw. For the first time in three billion years, rain would fall, and water would flow. And canals of our own design would carry water to irrigate the parched equatorial land.

Percival Lowell's vision of a lush Martian world populated by intelligent beings will have become reality. At long last, we will have found the Martians, and they will be us.

To the Stars

Someday, in the very distant future, we may break the bonds of our solar system, as we have already broken those of Earth, and take that epic leap beyond—to the stars.

In giant cosmic arcs, our descendants may begin a multi-trillion mile odyssey to discover worlds about which we can only dream today, to set down upon and explore distant lands, to colonize worlds throughout our galaxy.

This, then, may be the most important legacy we leave to our children, and to children for generations to come: the drive to explore other worlds—the will to better understand ourselves and our home planet.

And one day, we may finally realize the words of poet T.S. Eliot: "We shall not cease from exploration, and the end of all our exploring will be to arrive where we started, and to know the place for the first time."

To study the possibility of building self-contained spacecraft or colonies, and to learn about the Earth itself, scientists have developed the complex "Biosphere 2" in the desert near Tucson, Arizona.

Courtesy NASA

Courtesy NASA

THE HUBBLE TELESCOPE

The photographs on these two pages are all among the first taken by the crew of Discovery, *on April 30, 1990.* Far left: *The Hubble Space Telescope is shown here backdropped against Bolivia, Chile, and Peru. The remote manipulator arm of* Discovery *can be seen in the right side of the photo.* Left: *In this photograph, the Space Telescope is suspended by* Discovery's *remote manipulator system prior to the deployment of its solar panels and antennae. The arm can be seen in the right of the photo.* Below: *The Space Telescope is suspended in space following the deployment of its solar panels and antennae.*

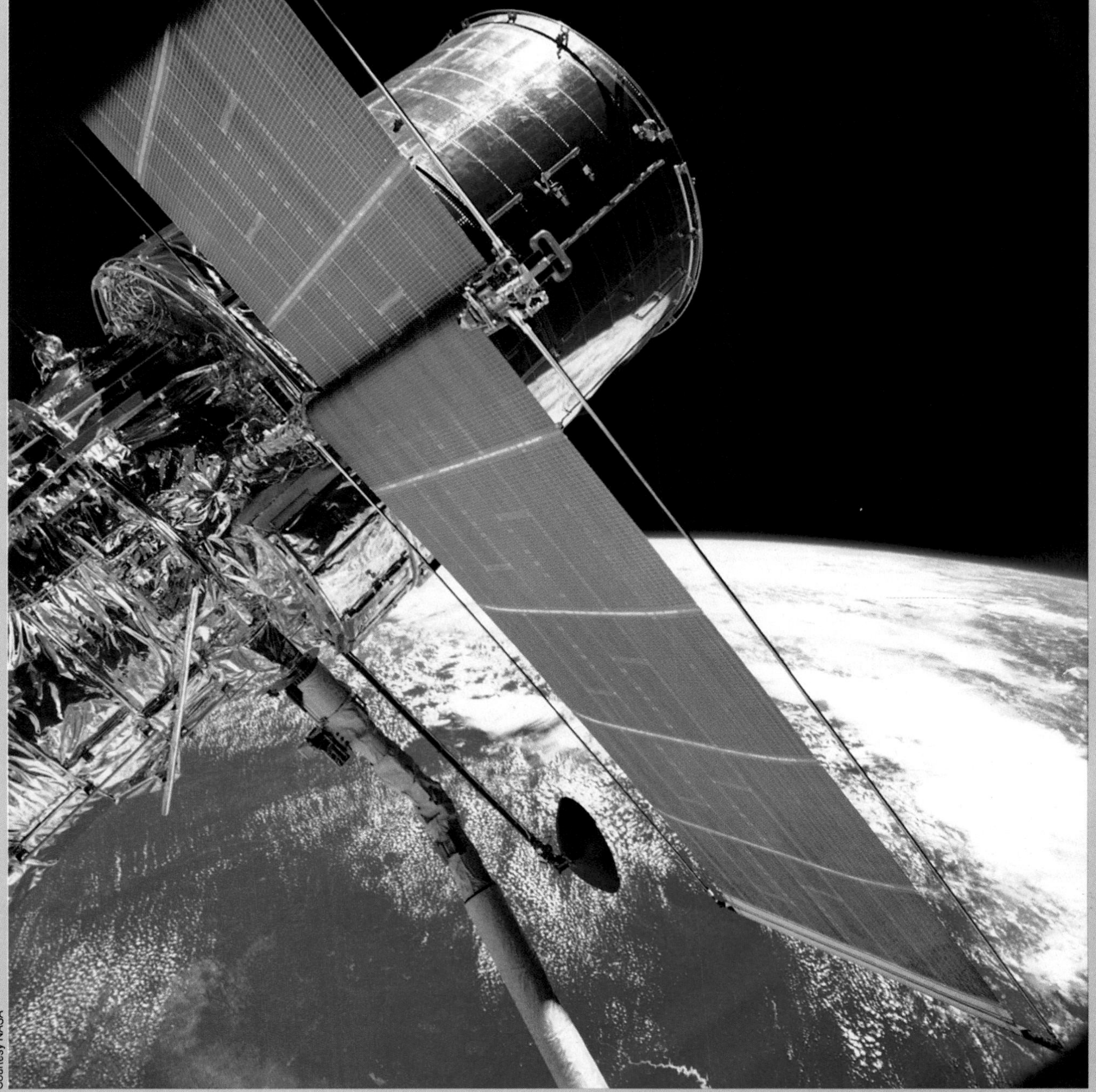

Courtesy NASA

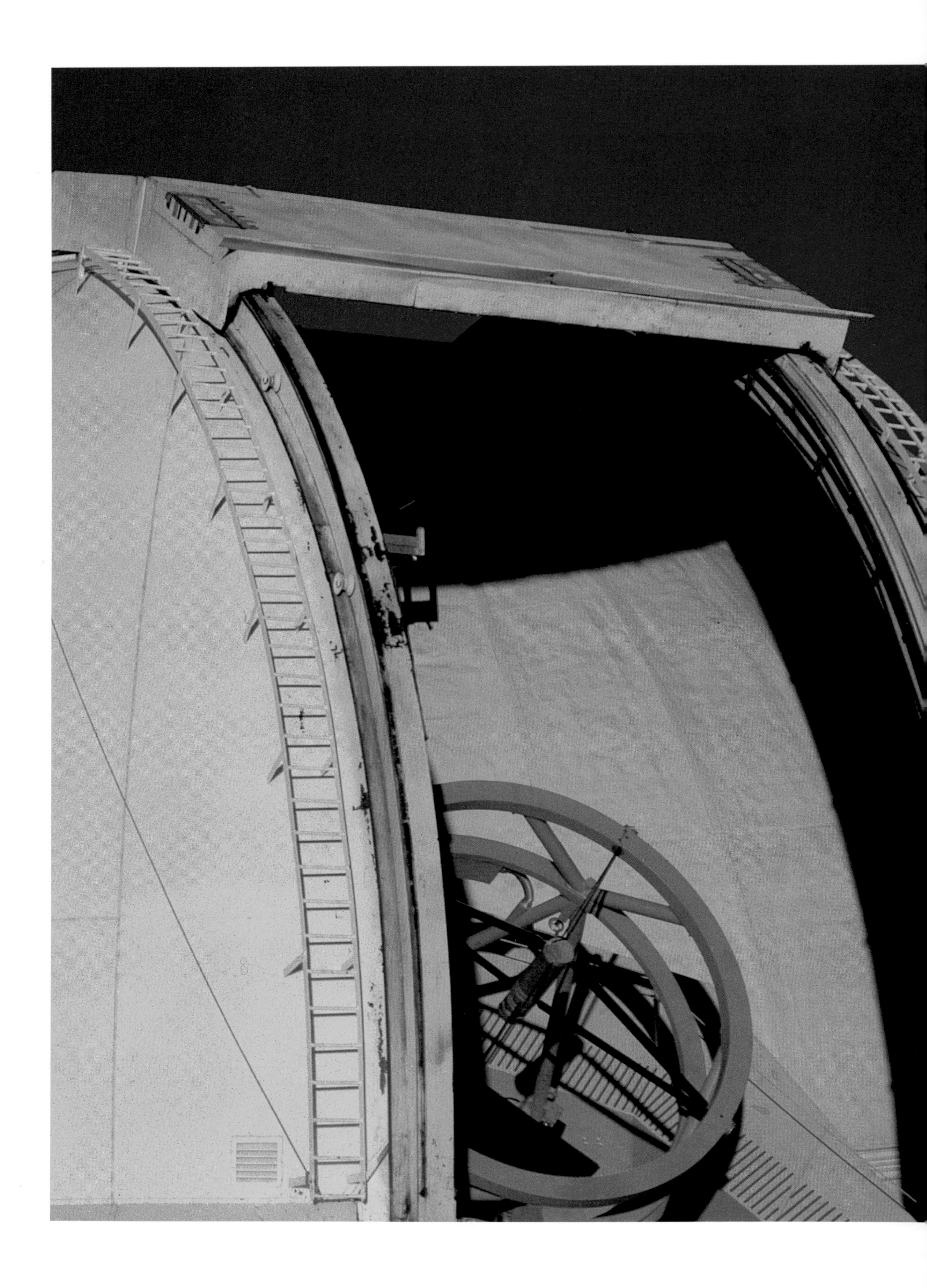

EPILOGUE

We of the latter half of the twentieth century have seen more than our ancestors could have ever dreamed. We have viewed up close eight of the nine known planets in our solar system, along with dozens of moons, and the icy nucleus of Halley's comet.

We have learned that our sun is but an average-sized star, tucked away among the clouds of gas and dust that form a spiral arm of the Milky Way galaxy. For decades, we have searched the stars of our galactic neighborhood for radio signals from intelligent beings. So far, we've found none.

We have seen that ours is not the only galaxy there is, but that scattered throughout the universe are hundreds of billions more of all shapes and sizes, each containing hundreds of billions of stars. We've learned that they cluster together in families that seem to form the surfaces of gigantic bubbles.

Near the edges of the universe we have found the enig-

With the huge infrared telescope on Mauna Kea in Hawaii, astronomers probe the mysterious realm of cosmic heat. They peer inside star-forming clouds, into the nuclei of galaxies, and search for other bodies much too cold and dark to glow in visible light.

matic quasars. Perhaps the cores of newborn galaxies, they may shine from the energy created as supermassive black holes gobble up a star the size of our sun every year.

And we have seen the remnant fires of the Big Bang that gave birth to our universe some fifteen or twenty billion years ago.

Courtesy Celestron International

Right: ***A region in the constellation of Orion, where new stars are continually being born. The Great Orion Nebula is one of the largest and most prolific star-forming regions of our galaxy.*** **Below:** ***The cracked lens from one of Galileo's original telescopes is mounted in an ornate ivory frame, and is preserved in the Museum of Science in Florence, Italy.***

In Search of our Roots

We have learned much about the universe in which we live. Yet there are those who ask what good it all is. Of what benefit is it to a starving, homeless family to study a quasar at the edge of the universe? Or to search for life among the stars? Or to land a man on the moon?

These are difficult questions at best, for often there is no immediate practical value from our research. There are, of course, countless spinoffs from technological developments that help improve our lives on Earth. But sometimes years, even centuries, must pass before the real value of our discoveries is known.

What we do know is that the study of our universe is, in a grand sense, the study of ourselves. The planet we live on, the air we breathe, the food we eat, even our own bodies, are made of chemicals forged in the cores of stars, and are blasted into space by cataclysmic explosions.

It is these chemicals which gave birth to life on our planet and, possibly, to life on countless other worlds throughout the universe. When we gaze at the stars on a clear, dark night, we are looking at our ultimate "roots."

Perspectives

Through the ages, the study of the cosmos has immeasurably influenced our view of life. Our language, art, literature, philosophy, religion, politics, and technology all reflect the perspectives we've gained from star hunters of ages past.

In fact, it's hard to imagine who we would be today had people not gazed upward in search of understanding. How would we think had we evolved on a cloud-covered planet—on a world where the sun, moon, stars, and planets, their cycles and mysteries, never appeared? What would have inspired our curiosity? Would we have even survived?

Our view of the cosmos has indeed changed over the ages. Yet, in many ways we are still ancient skywatchers gazing up toward the stars in search of answers.

How were the stars formed? Do other planets orbit the stars? How did galaxies form and evolve? Where did our universe come from and what is its fate? Are there other beings out there who share the cosmos with us?

As we approach the dawn of a new millenium, we stand on the threshold of an exciting age of discovery, one that will surely change how we view our universe and our place in it.

The star hunters are in for an exciting time.

From the time Galileo first trained a telescope at the moon, through the launching of the space telescope, advances in technology have led directly to a greater understanding of our universe.

ADDITIONAL PHOTO CREDITS

13(Bottom Right), 40-41, Giraudon/Art Resource,
22(Bottom Left), 23, 51(Top Right), 69(Bottom Right) Bridgeman/Art Resource
31(Top Right), 55(Top Right), 72(Bottom Left) Historical Picture Service/FPG International
32-33 © Gian Berto Vanni
34, 58(Bottom Left), 60(Bottom Left), 61, 65, 154(Bottom Left) Scala/Art Resource
35(Bottom Right) SEF/Art Resource
42(Bottom Left) Photoworld/FPG International
105 © Gary Ladd
8, 116, 127(Inset) Courtesy NASA
132 © David Mills/BMA The Photo Source
136-137 © Dan McCoy/Rainbow

INDEX